Principles of Radiometric Dating

The time dependent decay of naturally-occurring radioactive isotopes or in-growth of their radioactive or stable daughter products form the basis of radiometric dating of several natural processes. Developed in the beginning of the last century mainly to determine the absolute ages of rocks and minerals, radiometric chronology now plays a central role in a broad range of Earth and planetary sciences - from extra-solar-system processes to environmental geoscience. With the prerequisite of only college level knowledge in physics, chemistry and mathematics, this concise book focuses on the essential principles of radiometric dating in order to enable students and teachers belonging to diverse fields of studies to select, understand, and interpret radiometric dating results generated and published by professionals.

Kunchithapadam Gopalan is Honorary Scientist of the Indian National Science Academy. After his postdoctoral research on meteorites and moon rocks with Professor G. W. Wetherill at UCLA, Gopalan initiated modern isotope geoscientific studies in India at the Physical Research Laboratory, Ahmedabad and the National Geophysical Research Institute, Hyderabad. His notable distinctions include the Bhatnagar award of the Indian Council of Scientific and Industrial Research, M. S. Krishnan Gold Medal of the Indian Geophysical Union, Medal of Honour of the Indian Society for Mass Spectrometry, and Fellowship of the three national academies of science.

Principles of Radiometric Dating

Kunchithapadam Gopalan

University Printing House, Cambridge CB2 8BS, United Kingdom
One Liberty Plaza, 20th Floor, New York, NY 10006, USA
477 Williamstown Road, Port Melbourne, vic 3207, Australia
4843/24, 2nd Floor, Ansari Road, Daryaganj, Delhi – 110002, India
79 Anson Road, #06-04/06, Singapore 079906

Cambridge University Press is part of the University of Cambridge.

It furthers the University's mission by disseminating knowledge in the pursuit of education, learning and research at the highest international levels of excellence.

www.cambridge.org
Information on this title: www.cambridge.org/9781107198739

© Kunchithapadam Gopalan 2017

This publication is in copyright. Subject to statutory exception and to the provisions of relevant collective licensing agreements, no reproduction of any part may take place without the written permission of Cambridge University Press.

First published 2017

Printed in India by Nutech Print Services, New Delhi

A catalogue record for this publication is available from the British Library

ISBN 978-1-107-19873-9 Hardback

Additional resources for this publication at www.cambridge.org/9781107198739

Cambridge University Press has no responsibility for the persistence or accuracy of URLs for external or third-party internet websites referred to in this publication, and does not guarantee that any content on such websites is, or will remain, accurate or appropriate.

The author dedicates this book to the two authors of his being
(parents, Ramamrutham Kunchithapadam and Pattammal Kunchithapadam),
and the three authors of his becoming
(mentors, Prof V. S. Venkatasubramanian, Prof G. W. Wetherill and Prof D. Lal).

Contents

Preface — xi
Acknowledgments — xiii

1. Basics — 1

1.1 Nuclear Size and Constituents — 1
1.2 Fundamental Forces — 2
1.3 Nuclear Mass — 3
1.4 Equivalence of Mass and Energy — 3
1.5 Periodic Table — 4
1.6 Nuclear Composition and Stability — 6
1.7 Nuclear Binding Energy — 8
1.8 Cosmic Abundances — 9

2. Nuclear Transformations — 12

2.1 Introduction — 12
2.2 Spontaneous Nuclear Transformations — 12
2.3 Induced Nuclear Transformations — 15
2.4 Induced Nuclear Transformations in the Laboratory and Nature — 16
2.5 Role of Natural Radioactivity in Geodynamics and Geochronology — 17
2.6 Statistical Aspect of Radioactivity — 18
2.7 Binomial Distribution for Radioactive Disintegrations — 21
2.8 Poisson Distribution — 21

3. Nucleosynthesis — 23

3.1 Introduction — 23
3.2 Stellar Nucleosynthesis — 24

4. Isotopics — 33

4.1 Introduction — 33
4.2 Isotopic Abundance — 33
4.3 Isotope Effect in the Nuclear and Atomic Domains — 34
4.4 Notation of Isotopic Abundances — 34
4.5 Mixtures of Isotopically Different Components — 36

5. Radioactivity and Radiometric Dating — 40

- 5.1 Introduction — 40
- 5.2 Radioactive and Radiogenic Isotope Dating — 43
- 5.3 Long-lived Parent-daughter Couples used in Radiometric Dating — 48
- 5.4 Short-lived Parent-daughter Couples used in Radiometric Dating — 53
- 5.5 Interpretation of t, Radioactive Decay Interval — 58
- 5.6 Isochron Concept, Isotope Equilibration, and Closure Temperature — 58
- 5.7 Termination of a Single Stage Growth in the Past — 61
- 5.8 Recognition of Events Causing Isotopic Equilibration on Different Scales — 63
- 5.9 Projection of Present Day Isotopic Composition Back in Time — 64
- 5.10 Reservoir Ages — 66
- 5.11 Coupling Two Different, But Chemically Identical Parent-daughter Systems — 67
- 5.12 Concordia and Discordia — 71
- 5.13 Coupling Two Chemically Different Decay Systems — 74
- 5.14 Chemical and Half-life Diversity of Parent-daughter Pairs — 76
- 5.15 Radiometric Dating by Indirect Radiogenic Effects — 77
- 5.16 Conclusion — 79

6. Mass Spectrometry and Isotope Geochemistry — 80

- 6.1 Introduction — 80
- 6.2 Principles of Mass Spectrometry — 81
- 6.3 Ion Detectors — 82
- 6.4 Sequential vs Simultaneous Detection of Ion Beams — 83
- 6.5 Improved Mass Spectrometers — 85
- 6.6 Types of Ion Sources Used in Isotope Geochemistry — 86
- 6.7 Typical Commercial Mass Spectrometers Using Different Ion Sources — 88
- 6.8 Mass Fractionation in Mass Spectrometers — 91
- 6.9 Absolute Abundance of an Isotope — 92
- 6.10 Sample Size Requirements — 93
- 6.11 Mass Spectrometry vs Decay Counting — 93
- 6.12 Accelerator Mass Spectrometer — 94

7. Error Analysis — 96

- 7.1 Introduction — 96
- 7.2 Systematic and Random Errors — 97
- 7.3 Measurement of Random Data — 97
- 7.4 Population Mean and Sample Mean — 98
- 7.5 Propagation of Measurement Uncertainties — 101
- 7.6 Standard Deviation of the Mean of n Measurements — 102
- 7.7 Joint Variation of Two or More Random Variables — 103
- 7.8 Regression Analysis — 105
- 7.9 York's Solution — 106
- 7.10 Measure of Goodness-of-fit — 109

8. Meteorites: Link between Cosmo- and Geochemistry — 110
- 8.1 Introduction — 110
- 8.2 Nucleocosmochronology — 111
- 8.3 Extinct Nuclides and Formation Interval — 112
- 8.4 Meteorites — 114
- 8.5 Nebular condensation — 117
- 8.6 Planetary Accretion — 118
- 8.7 Isotope Abundances in the Solar Nebula — 119

9. Chronology of Meteorite History — 121
- 9.1 Introduction — 121
- 9.2 Stage 1: Formation Intervals from Extinct Isotopes — 124
- 9.3 Stages 2 through 4: Formation Ages of Meteorites — 127
- 9.4 Rb-Sr, Sm-Nd and U-Pb Ages of Meteorites — 129
- 9.5 Very High Precision Model Ages — 131
- 9.6 Meteorite Ages Much Younger than 4.5 Ga — 134
- 9.7 Stage 5: Gas Retention Ages and Post-Formational Cooling and Heating Histories — 134
- 9.8 Stage 6: Duration of Meteorites as Small Independent Objects in Space — 137
- 9.9 Stage 7: Terrestrial Residence Time of Meteorite Finds — 139

10. Chemical Evolution of the Earth — 140
- 10.1 Composition of Terrestrial Planets and Chondritic Meteorites — 140
- 10.2 Energetic Processes During the Final Stages of Earth Accretion — 141
- 10.3 Element Segregation: Some Geochemical Rules — 142
- 10.4 Segregation of Major and Trace Elements During Melting or Igneous Processes — 143
- 10.5 Graphical Representation of Inter-Element Variations in Compatibility — 145
- 10.6 Melting and Crystallization Models — 146
- 10.7 Combined Partial Melting and Recrystallization — 150
- 10.8 Observational Constraints on the Structure and Composition of the Modern Mantle — 151
- 10.9 Earth as a Large Geochemical System — 155
- 10.10 Elemental Chemistry of Mid-Ocean-Ridge Basalts, Ocean-Island Basalts and Continental Crust — 157

11. Chronology of Earth History — 161
- 11.1 Introduction — 161
- 11.2 Early Siderophile-Lithophile Segregation and Timing of Core Formation — 163
- 11.3 Early Lithophile-Atmophile Separation and Timing of the Primitive Atmosphere — 165
- 11.4 Lithophile-Lithophile Separation and Timing of the Early Crust — 168
- 11.5 ^{142}Nd Evolution in the Earth's Mantle — 171
- 11.6 ^{143}Nd Evolution in the Earth's Mantle — 172
- 11.7 ^{87}Sr Evolution in the Earth's Mantle — 175
- 11.8 Coupling Neodymium and Strontium Data — 176
- 11.9 ^{206}Pb, ^{207}Pb Evolution in the Earth's Mantle — 180

11.10 Evolution of ^{176}Hf in the Earth's Mantle ... 182
11.11 Evolution of ^{187}Os in the Earth's Mantle ... 182
11.12 Evolution of Strontium and Neodymium Isotope Ratios in Seawater ... 183
11.13 Magma Sources in the Mantle ... 185
11.14 Evolution of Radioactive Daughter Isotopes ... 188
11.15 Giant Impact Hypothesis ... 191

References ... *192*

Index ... *203*

Preface

The time-dependent accumulation of helium and lead from the radioactive decay of uranium in minerals and rocks was suggested by Rutherford in 1905 as a means of determining their absolute ages. This seminal idea has been assiduously pursued in the last century by unorthodox physicists and chemists to detect and quantitatively measure numerous radioactive isotopes of widely varying lifetimes and abundances in natural systems. Absolute age determination based on these isotopes, called radiometric dating, now plays a central role in a broad range of Earth and planetary sciences: paleoseismology; paleomagnetism; paleooceanography; igneous, metamorphic and sedimentary petrology; geomorphology; geochemistry; tectonics; nucleosynthesis; cosmochemistry; planetary science; geobiology; paleoclimatology; paleoanthropology; and archeology. Assuming that the reader has only college level knowledge of physics, chemistry, and mathematics, this concise book (about 200 pages) focuses on the essential principles of radiometric dating in order to enable the students and teachers in various fields to quickly figure out the criteria to be met by parent-daughter systems and samples relevant to their specialization. This book draws heavily on three classic review articles which, in my view, capture the intellectual appeal and beauty of the subject for a very wide audience (Wetherill et al., 1981; Wasserburg, 1987; Allegre, 1987). I believe that this book will succeed in improving students' understanding and appreciation of radiometric dating results generated and published by professionals. I hope it would also stimulate interest in students to take up isotope geology as a serious study and reach out for the excellent and comprehensive books on the subject.

The material presented in each of the 11 chapters is self-contained. However, the reader is urged to read all the chapters, as they are strung together into a concise, continuous, and easily comprehensible narrative to illuminate the subject as a whole. Vital points behind radiogenic isotope chronometry are stressed upon more than once. The reference list is mainly for students interested in further reading.

Chapter 1 covers the basic facts of nuclear and atomic physics, nuclear binding energy as a measure of nuclear stability, and the variety and relative abundance of different elements in the sun and the primitive meteorites. Chapter 2, then, moves on to the transformation of composition of nuclides, either spontaneously (radioactivity) or by external agents (induced nuclear reactions), and highlights the role of feeble natural radioactivity, both in driving and dating planetary processes. It ends with a section on an important aspect of radioactivity, namely its statistical nature. Chapter 3, on nucleosynthesis, builds on the first two chapters to show that the process of formation of elements (strictly-speaking, stable and unstable nuclides) inherited by the Solar System, is due to a complex combination of nuclear reactions induced by charged particles in the interior of stars, and by neutrons in the final stages of stellar evolution. The chapter brings out the important point that the science of radiometric chronology, pivotal in many discrete disciplines, rests on the time required by various unstable nuclides, produced in stellar and terrestrial nuclear reactions, to reach a stable nuclear composition. Chapter 4 on 'isotopics' introduces the reader to the practical unit of measuring isotope abundances, the dramatic differences in isotope effects between nuclear and atomic domains, and familiarizes the reader with the notation of isotopic mixtures of an element. The chapter ends with the simple mathematical derivations of mixtures of isotopically different elements.

There are many and somewhat different ways in which the radioactive isotopes and their daughter products are used to date natural events and processes. But, the one common feature of these different methods is that they all measure a radioactive decay interval. It is only the meaning and interpretation of a measured decay interval that depends on the scientific context, the sample analyzed, the decay system selected, and the analytical method employed. Chapter 5 builds on this underlying unifying principle to show that all known applications of radiometric dating follow from the creative use of the basic radioactive decay process and imaginative selection of natural samples. I believe that this compact and generalized treatment can enhance critical and individualistic thinking among users of radiometric age results. The mathematical equations in this chapter are difficult only superficially, as they are based on elementary mathematics.

Advancements in mass spectrometers have largely dictated the advances in radiometric dating. Chapter 6 relies on simple figures to explain the three basic components of mass spectrometers and improvements in each of them, in a reader friendly way. Chapter 7 provides a rigorous yet easily understandable treatment of statistical error analysis.

Published text typically examines each isotope system separately, chapter after chapter. In contrast, chapters 8, 9, 10, and 11 in this book explore a few aspects of the evolutionary chronology of meteorites, the least evolved planetary objects available for laboratory analysis, and the Earth, a highly evolved planet, respectively, by examining relevant isotopic systems that illuminate the temporal aspects of evolution. Illustrations or case studies have been carefully chosen to bring out the excitement of research and discovery in radiometric chronology, in particular, and radiogenic isotope geology, in general.

Finally, each chapter begins with an apt quotation(s) from famous personalities to alert the readers of its content and also to leave a lasting impression on their minds.

Readers are invited to write to me, at *gopalank1@rediffmail.com*, regarding any errors, comments, and suggestions for improvement. Positive feedbacks are, of course, welcome, if they are warranted. I do hope that readers from various fields will find this book a good investment as well as a quick reference source.

Acknowledgments

I am grateful to Professor J. D. Macdougall and Professor M. E. Bickford for reading all the chapters of the book in order to provide constructive suggestions, improvements in its presentation, and the correction of numerous errors in the language. I am indebted to Professor R. K. O'Nions and Professor Bor-ming Jahn for their encouragement, constructive criticism and suggestions for improvement. However, all the shortcomings in this book and the errors of omission are entirely my responsibility. I thank V. Rajasekhar for the art work. I thank M. Choudhary of Cambridge University Press in India for his encouragement and taking the necessary decisions that enabled me to publish this book. I am very grateful to the Indian National Science Academy for giving me the position of the Honorary Scientist, which enabled me to write this book. Finally, my thanks go to my wife, Savithri and son, Ramesh for their understanding and support.

I am grateful to the publishers and authors listed below, for permission to copy or redraw figures from their journals and books for which they hold the copyrights. Reference to the original source of such figures is given within individual figure captions.

Academic Press, UK	
Using geochemical data (Rollinson, 1993)	2 figures
American Association for the Advancement of Science	
Science	1 figure
American Astronomical Society	
Astrophys. J (lett)	1 figure
American Physical Society	
Physical Review Letters	1 figure
Annual Reviews, Inc	
Annual Reviews of Earth and Planetary Sciences	1 figure
Cambridge University Press	
Radiogenic isotope geology (Dickin, 2005)	1 figure
The solid earth (Fowler, 1990)	1 figure
Geochemsitry: An Introduction (Albarede, 2003)	1 figure

Elsevier Science Publishers
- *Geochimica et Cosmochimica Acta* — 2 figures
- *Earth and Plaetary Science Letters* — 1 figure
- *Earth as an evolving planetary system (Condie, 2005)* — 1 figure

Indian Society for Applied Geochemists
- *J. Appl. Geochemistry* — 12 figures

John Wiley Publishers
- *Geochemistry (White 2013)* — 2 figures
- *Early Earth Systems (Rollinson, 2007)* — 2 figures

Harvard University Press
- *From Stone to star (Allegre, 1992)*

Nature Publishing Group
- *Nature* — 3 figures

Prentice Hall
- *Principles of igneous and metamorphic petrology (Philpotts, 1990)* — 2 figures
- *Geochemsitry: Pathways ad processes (Richardson and McSween, 1989)* — 1 figure

Princeton University Press
- *How to build a habitable planet (Langmuir and Broecker, 2012)* — 4 figures

Royal Society of London
- *Philosophical Transactions, Royal Society of London* — 1 figure

Springer-Verlag GmbH
- *Absolute age determination (Geyh and Schleicher, 1990)* — 1 figure

Stanford University Press
- *The age of the earth (Dalrymple, 1991)* — 1 figure

Taylor and Francis group
- *Thinking like a physicist (Thompson, 1990)*

University science books
- *The Physical Universe: An introduction to Astronomy (Shu, 1982)* — 1 figure

1 Basics

> The standard model of particle physics represents our deepest knowledge about what the world is made of. But this theory is far from unique and does not explain why the proton, neutron and electron masses are so finely tuned.
>
> **Lee Smolin**

1.1 Nuclear Size and Constituents

It is now well known that atoms of chemical elements consist of a nucleus surrounded by a cloud of electrons. Although it is generally appreciated that an atom is a very small entity, quantitative figures like ~10^{-10} m for its size are not easy to grasp. In Science, very large and small decimal numbers are conveniently expressed in terms of powers of 10, known as scientific notation. Some examples, spanning 36 orders of magnitude are given in the Table 1.1. Prefixes like milli, micro, and kilo may be familiar. It is useful to become familiar with the meaning of other prefixes. The drawback of scientific notation is that it desensitizes us to the enormous range of numbers that the scale of the universe demands. For example, the famous scientist Rutherford did an elegant and simple experiment to show that the size of the nucleus is not comparable to that of the atom, but about five orders of magnitude smaller: ~10^{-15} m. The difference in size given in scientific notation may not impress us until we realize that if the atom is magnified to the size of a cathedral, the nucleus will be no bigger than a housefly.

Table 1.1 Decimal Fractions and Multiples

Fraction	Prefix	Symbol	Multiple	Prefix	Symbol
10^{-3}	milli	m	10^{3}	kilo	k
10^{-6}	micro	μ	10^{6}	mega	M
10^{-9}	nano	n	10^{9}	giga	G
10^{-12}	pico	p	10^{12}	tera	T
10^{-15}	femto	f	10^{15}	peta	P
10^{-18}	atto	a	10^{18}	exa	E

The nucleus consists of two distinct particles: protons with an electrical charge (conventionally taken as positive), and neutrons with no net electrical charge. Protons and neutrons are collectively known as nucleons. A combination of protons and neutrons, stable or otherwise, is called, in general, a nuclide. Electrons have an equal and opposite electrical charge (taken as negative) to protons. In an electrically neutral atom, the number of protons in the nucleus is equal to that of electrons orbiting it. This is known as the proton number when referring to the nucleus, and atomic number (Z) when referring to the atom as a whole. If an atom loses an electron, it will have a net positive charge and is called a positive ion, or cation; if it gains an electron, it will have a net negative charge and is called a negative ion, or anion.

1.2 Fundamental Forces

Electrons in an atom are bound to their nucleus by the electromagnetic force, which is attractive between unlike charges and repulsive between like charges. This raises the question as to how the protons are bound within the small nuclear volume despite their mutual repulsion. The answer is that a much stronger force, known as the strong or nuclear force, acts between nucleons. But as it is effective only within nuclear dimensions, two protons must first be brought close together against their electrostatic, or Coulomb repulsion, before the strong attractive force can hold them together. There is a third force, the weak nuclear force, with an effective range even smaller than that of the strong nuclear force. The only other force known in nature is the familiar gravitational force, which is many orders of magnitude weaker than the other three. Table 1.2 shows the differences in the strength and range of the four forces. The strong and weak forces do not extend beyond nuclear dimensions, whereas, the electromagnetic and gravitational forces reach out to the farthest corners of the universe. The four known forces separated and became individually distinct during the earliest moments of the universe (Chapter 3).

It was noted above that the strong force between protons balances their mutual electrostatic repulsion within nuclear dimensions. It should, therefore, be possible to form nuclei containing only protons. But all elements heavier than the lightest hydrogen contain protons and neutrons, in comparable number, in their nucleus. This implies that a combination of protons alone is not stable, and the pure attractive short-range nuclear force between protons and neutrons is additionally required for nuclear stability. It is also found that the number of neutrons for a given number of protons can vary slightly. Nuclides with the same number of protons, but different number of neutrons are called isotopes. The sum of the proton number (Z) and neutron number (N) is called mass number (A). Nuclides with the same number of neutrons, but different number of protons are called isotones, and nuclides with the same number of nucleons (A) are called isobars.

The symbol of an atom is just its chemical symbol without any explicit reference to its atomic number (Z) and mass number (A). For example, Cl for chlorine and Sm for samarium. The symbol for an isotope must necessarily include its mass number, usually as the left superscript. For example, ^{35}Cl and ^{147}Sm. The atomic number is understood from the chemical symbol. The use of left superscript to designate the mass number avoids the confusion with exponents, and retains the option to define the right superscript and subscript, as needed.

Basics

Table 1.2 Characteristics of the Four Fundamental Forces

Force	Source	Strength	Range (cm)
Strong	Baryons, mesons	1	10^{-13}
Electromagnetic	Photons	10^{-2}	Infinity
Weak	Leptons, mesons	10^{-5}	10^{-15}
Gravitational	Mass/energy	10^{-40}	Infinity

1.3 Nuclear Mass

Like their sizes, masses of atoms are also extremely small: about 27 orders of magnitude smaller than the standard mass unit, the kilogram. As masses of individual atoms figure prominently in nuclear, atomic, and molecular physics, they are measured in a more convenient small mass unit, called the atomic mass unit (amu) or simply the unified mass unit (u). It is defined as a twelfth of the mass of the ^{12}C atom, and is equal to $1.6605389 \times 10^{-27}$ kg.

Mass of macroscopic objects is measured with analytical balances on the basis of the gravitational force on such objects. The smallest mass that can be determined in sophisticated analytical balances is about one microgram (10^{-9} kg), which is much greater than that of individual atoms. The analytical instrument commonly used to measure atomic masses is called a mass spectrometer (Chapter 6), which is based on the electromagnetic force (that is very much stronger than the gravitational force, but follows the same inverse square dependence on distance) on moving electrically charged atoms (ions). Mass spectrometers measure atomic masses relative to that of ^{12}C atom (12 u). The values of atomic masses reported in the literature are not nuclear, but isotopic masses; they include the mass of the extra nuclear electrons in neutral atoms. This convention has some advantages in the treatment of nuclear reactions and energy relations. Atomic masses have now been determined with precisions varying between one and 10 parts per million; a few examples are given in the Table 1.3.

Table 1.3 Nucleon and Nuclear Masses

Particle	Symbol	Mass (u)
Proton	p	1.007276
Neutron	n	1.008665
Electron	e^-	0.000549
Hydrogen	^1H	1.007825
Helium	^4He	4.001475
Carbon	^{12}C	12.00000

1.4 Equivalence of Mass and Energy

One of the important consequences of Einstein's Special Theory of Relativity is the equivalence of mass and energy. The total energy content E of a system of mass M is given by the relation, $E = Mc^2$, where c is the velocity of light in vacuum (2.99976×10^{10} cm.s^{-1}). The energy equivalent of 1u is:

$$E = 1.661 \times 10^{-27} \text{ Kg} \times (2.999 \times 10^{10} \text{ cm})^2 = 1.493 \times 10^{-23} \text{ erg}$$

An energy unit much more convenient in nuclear work than the erg is the electron volt (eV), the kilo electron volt (KeV), and the million electron volt (MeV). The electron volt is defined as the energy necessary to raise one electron through a potential difference of one volt, akin to the increase of potential energy of a mass when raised in the gravitational field of the earth. 1 eV = 1.662×10^{-12} erg, which gives 1 u = 931 MeV. MeV is a large energy in the nuclear scale, but extremely small in the human scale. The kinetic energy of a small ant moving at a speed of a few mm per second is a few thousand MeVs.

Table 1.4 Mass-energy Equivalence of Protons, Neutrons, and Electrons

Particle	Mass (u)	Energy (MeV)	Electric Charge (Coulomb)
Proton	1.007276	938.3	$+1.60 \times 10^{-19}$
Neutron	1.008665	939.6	0
Electron	0.000549	0.51	-1.60×10^{-19}

Table 1.4 shows that almost the whole mass of an atom is concentrated in its nucleus even though it accounts for only a trillionth of the total volume of the atom. The extra nuclear electrons contribute very little to the total mass, but represent the volume of the atom.

1.5 PERIODIC TABLE

When Mendeleyev developed his famous periodic table of elements in 1871, the structure of atoms was not known. So he organized the elements known in his time according to their atomic mass. It is now known that the chemical properties of elements depend very little on the composition, and, hence, the mass of their nuclei, but mainly on the configuration of the extra-nuclear electrons. The modern periodic table is based on the electronic structure described by quantum mechanics. According to quantum mechanical principles, each electron in a multi-electron atom is characterized by a unique set of four quantum numbers, namely principal, azimuthal, magnetic, and spin quantum numbers. The meaning and role of each quantum number is given in many text books (Aitkins, 1986). For our purpose, it is sufficient to consider that electrons are organized in 'shells'. The shells can be grouped in broad categories called the first shell (also known as the K shell), the second (or the L shell) and so on. They are given numbers 1, 2, 3, 4, etc. Each such major shell contains subshells: only one subshell ($1s$) in the first shell, two subshells ($2s$, $2p$) in the second, three subshells ($3s$, $3p$, $3d$) in the third shell, four subshells ($4s$, $4p$, $4d$, $4f$) in the fourth, and so on. There is a limit to the maximum number of electrons that can be accommodated in any one of the subshells: 2 in s-type, 6 in p-type, 10 in d-type, and 14 in f-type. The energy levels of the subshells vary. The order of increasing energy for the subshells or orbitals is shown in Figure 1.1.

The diagram in Figure 1.1 is qualitatively correct for almost every neutral atom, and can be used to find the electron configuration of all, but a few elements in the periodic table shown in Figure 1.2. The modern periodic table in its long form consists of 18 vertical sequences of elements known as 'groups' and horizontal sequences called 'periods' (Mahan, 1975). Each period starts with an element which has one electron in an s-orbital. The first period is only two elements (H and He) long, since the $1s$-orbital or subshell can accommodate only two electrons. The third electron in lithium (Li) must enter the $2s$-orbital, and the second period begins. Since the $2s$ and $2p$ orbitals or subshells of the second shell can accommodate a total of eight (2+6) electrons, eight elements enter the table before the $2s$ and $2p$ orbitals are filled in the element neon. The third period is also eight elements long and ends when the $3s$ and $3p$ orbitals or subshells are filled in argon.

Basics

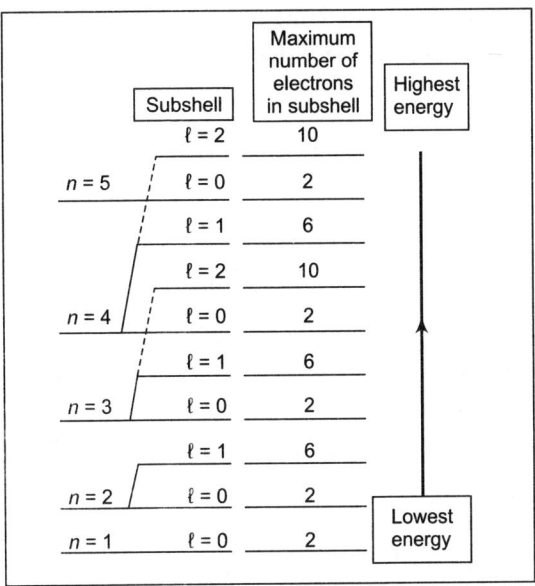

Figure 1.1 Pattern of orbital energies for neutral atoms.

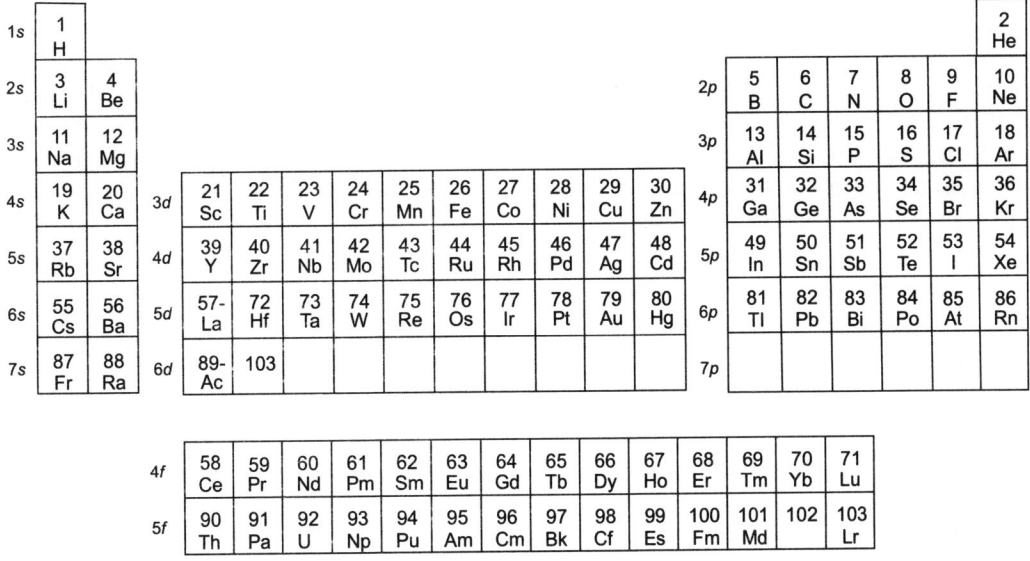

Figure 1.2 The periodic table showing the separation into the *s*-, *p*-, *d*-, and *f*-blocks.

Since the 4*s*-orbital is lower in energy than the 3*d*-orbital, a new period starts with potassium before an electron enters the 3*d*-orbitals. After the 4*s*-orbital is filled with two electrons in calcium (Ca), the five 3*d*-orbitals are the next available in order of increasing energy. As these 3*d*-orbitals can accommodate 10 electrons, the so-called 10 transition elements enter the table at this point. Once these 10 elements have entered, the fourth period is completed by filling the 4*p* orbitals with six electrons in krypton. In the fifth period, the 5s-, 4p- and 5p-orbitals are filled in succession, as in the fourth period. The sixth period is different in that after the 6s-orbital is filled, and one 5d-electron enters to form the element lanthanum, the 4f-orbitals are the next available in order of increasing energy. Since the f-orbitals can accommodate a total of 14 electrons, 14 elements enter the table before any of the remaining 5d-orbitals can be filled in. After the 14 so-called rare earth elements have entered the table, the last set of transition metals appears as the 5d-orbitals are occupied. These in turn are followed by six elements required to fill the 6p-orbitals, and the sixth period ends with radon. The seventh period starts by filling the 7s-orbital and after one 6d-electron appears, subsequent electrons enter the 5f-orbitals. Thus, the periodic table ends with the actinide series, a group of 14 elements analogous in properties and electronic structure to the rare earths. To keep the periodic table from becoming excessively long, the elements which follow lanthanum and 14 elements after actinides are placed in separate rows at the bottom of the table. Figure 1.2 also shows that the elements in which s-, p-, d-, and f-orbitals are being filled are grouped naturally in the long form of the table. The eight families of the s and p blocks are often called representative elements and those in the d bock are called transition elements, while members of the f block are known as the inner transition elements.

Group 18 elements are the so called noble gases. Their unreactive nature is due to the high stability arising from full s- and p-orbitals in their outermost shell. Group 1 elements have a single valence electron outside the noble gas configuration and exhibit metallic properties. Group 17 elements have one electron short of noble gas configuration and precede the noble gases. They exhibit nonmetallic properties (halogens). Elements in groups from 2 to 17 show progressive gradation of properties between the two extreme metallic and non-metallic groups.

The elements in the same column of the periodic table have, for the most part, the same configuration for their valence (outermost) electrons, and, as is well known, they have similar chemical properties. Horizontal 'chemical similarity' also exists, such as among the rare earth and transition metals. These elements, which are chemically similar, differ only by the number of electrons in a particular type of orbital, such as 3d and 4f. In addition to these general correlations between electron configurations and chemical properties, there are many more detailed correlations which are of direct interest to chemists and geochemists.

The periodic table shows how the electronic structure of elements almost exclusively governs their chemical properties. But electrons can be configured only around bare nuclides with a stable or metastable combination of protons and neutrons. The elements technetium (Tc) ($Z = 43$) and promethium (Pm) ($Z = 61$), and transuranic elements ($Z = 93$ and above) do not occur naturally at present on the earth. This implies the absence of nuclides with these proton numbers. When bare transuranic nuclides were artificially produced, they readily attracted the appropriate number of electrons from the surrounding to form the remaining elements of the actinide group. As each isotope of a natural element requires a specific nuclide composition, the number of different nuclides to form the known 90 chemical elements is much more, about 265.

1.6 Nuclear Composition and Stability

The distribution of protons and neutrons in the stable, or nearly stable, 265 nuclides is shown in Figure 1.3, known as the chart of nuclides (Langmuir and Broecker, 2012). In this chart horizontal rows of nuclides have the same Z, but different N to represent the different isotopes of a single element. Vertical lines of nuclides have the same N, but different Z known as isotones. Diagonal lines contain nuclides with the same nucleon number, known as isobars. The 265 stable nuclides conform to a narrow band across the chart, called, appropriately, the path of stability. Nuclides with low atomic numbers have approximately equal numbers of protons and neutrons, while heavier nuclides have many more neutrons than protons to stabilize them against the increasing electrostatic repulsion between protons confined to the small nuclear volume. The heaviest stable nuclide has $Z = 83$ and $N = 126$ corresponding to the isotope ^{209}Bi of bismuth. Nuclei heavier than ^{209}Bi are not stable with time and undergo radioactive decay (Chapter 2), but nuclides with $Z = 90$ (thorium) and $Z = 92$ (uranium), respectively, have extremely long lifetimes (Chapter 3) to survive to the present day.

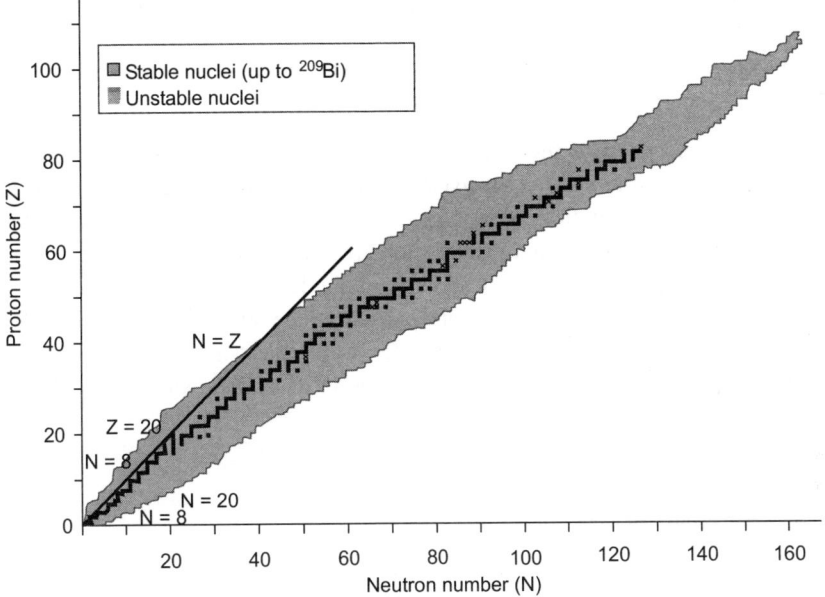

Figure 1.3 The chart of the nuclides. Stable nuclides (the band of stability) are indicated by the solid black boxes. Radioactive nuclides that decay back to the band of stability at variable rates are indicted by the gray field (Langmuir and Broecker, 2012)

A nuclear origin for the relative stability among the 265 nuclides is indicated by the far higher number of nuclides with even Z and N, as given in Table 1.5.

Table 1.5 Distribution of Stable Nuclides with Odd/Even *A*, *Z*, and *N*

A	Z	N	Stable Nuclides	Long-lived Nuclides
Even	Even	Even	161	11
Odd	Even	Odd	55	3
Odd	Odd	Even	50	3
Even	Odd	Odd	4	5

More than half (160) of the stable nuclides have even Z and even N. Stable nuclides with even Z and odd N, or vice versa are much less common (~50). In great contrast, stable nuclides with both Z and N odd are extremely rare (4), namely ^{2}H of hydrogen, ^{6}Li, ^{10}B of boron, and ^{14}N of nitrogen. Why are nuclides with even Z and even N more stable and hence abundant, as shown above? Whenever nucleons, either protons or neutrons, can pair with their spins in opposite directions, the forces holding them seem to be stronger and the nuclide more stable. This is very similar to the stability of atoms with paired electrons in each orbital. In fact, similar to the much higher stability (chemical inertness) of noble gas atoms with filled orbitals, nuclides with particular numbers (2, 8, 20, 28, 50, 82, 126) of either Z or N are exceptionally stable. These are called magic numbers.

A very large number of nuclides (>2,000) have been artificially produced in laboratories. All these nuclides fall away on either side of, or beyond, ^{209}Bi along the path of stability, and invariably are unstable. Given sufficient time, all unstable nuclides (both natural and artificial) will transform or decay in one or more steps to stable nuclides in the path of stability. This spontaneous time-dependent process is called radioactivity (Chapter 2), and the unstable nuclides are called radioactive nuclides, radionuclides, or radioisotopes.

The presence, at present, of a few long-lived radionuclides, like ^{238}U of uranium, in nature suggests that some of the relatively short-lived nuclides (on either side of the path of stability) might also have been produced with them at some time in the past, but have all decayed to extinction since. The present occurrence of some of these short-lived nuclides at present must then be due to their continuous production, either from the series or chain decay of long-lived nuclides, or nuclear reactions due to cosmic rays or their secondary particles. The still surviving long-lived nuclides, short-lived nuclides supported by the former, and a few of the short-lived nuclides that have become extinct in materials available to us at present form the basis of the extensive field of radiogenic isotope geology.

1.7 NUCLEAR BINDING ENERGY

The general principle that the lowest energy configuration of an isolated system (large or small) is the most stable accounts for the stability of some combinations of nucleons relative to a system of the same number of free or separate nucleons. We can get some insight into the energy with which nucleons are bound to a nucleus like helium (with two protons and two neutrons) as follows; Nuclear Binding Energy (BE) is the energy required to separate a nuclide into its constituent nucleons and, hence, a measure of its stability.

Mass of ^{4}He: 4.0026 u

Mass of 2p + 2n + 2e: 2.04652 + 2.01733 + 0.001098 = 4.03298 u

Mass difference (mass defect) = 0.00304 u or 28.3 MeV

Binding Energy (BE) per nucleon = 28.3/4 = 7.1 MeV

The BE per nucleon vs mass number is shown in Figure 1.4. The distinct peak at ^{4}He of helium (7.1 MeV) represents a very stable nuclear package. The BE increases rapidly from ~1 MeV for low masses to a maximum of 8.7 MeV for A ~60, and decreases smoothly, thereafter. The BE is insufficient to stabilize nuclei heavier than ^{209}Bi. They decay spontaneously to rearrange their nucleon composition.

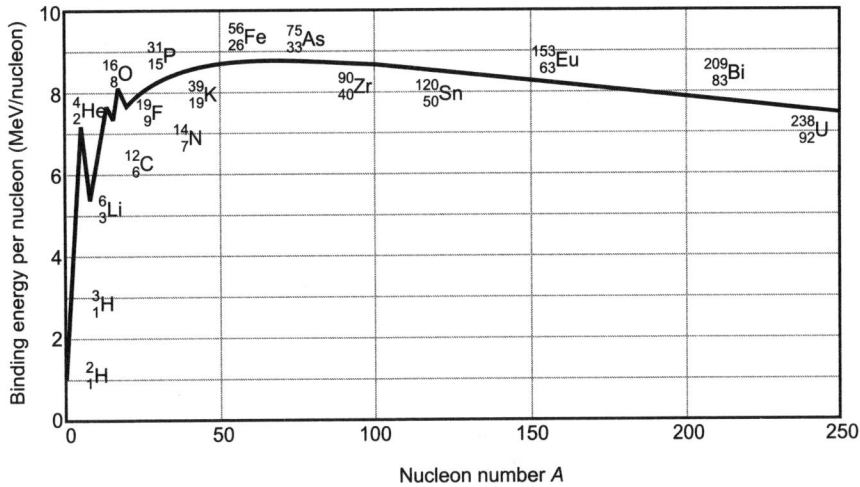

Figure 1.4 Binding energy/nucleon vs mass number diagram

We have so far seen that naturally occurring elements can be systematized, both in terms of their chemical properties (periodic table) and their nuclear attributes (chart of nuclides). We have not considered how, where, and when they were produced in the first place and their relative abundances in the pristine matter from which the solar system formed. We will consider the latter first and defer the former to a later chapter.

1.8 Cosmic Abundances

Our solar system is believed to have resulted from the gravitational collapse of a large cloud of gas and fine dust from the interstellar medium. As the sun is the bulk repository of this cloud of matter, its composition must be representative of that of the parent cloud. The white light emitted by the sun is absorbed at characteristic wavelengths by the chemical elements in its outer atmosphere (photosphere) in proportion to their relative abundances. Spectroscopic measurements of such absorption bands in the otherwise continuous solar spectrum can be converted to relative elemental abundances, but not with high precision and accuracy. Elemental abundances of distant stars in other galaxies are also obtained by this remote sensing method. By convention, the relative abundance of an element is stated as the number of atoms of that element for each one million atoms of silicon (Si). A tabulation of solar abundances is given by Ross and Aller (1976). Relative abundances in the photosphere (surficial part of the sun) are taken as representative of the embryonic sun, as nuclear reactions taking place in the core of the sun (Chapter 3) are believed not to affect the original photospheric composition.

A tangible sample of the pristine solar system material can be directly and precisely analyzed in a laboratory for relative abundances of both elements and their isotopes. Tangible samples from the surficial parts of the earth and even the moon can hardly be representative of the whole earth or the moon, much less of the primitive solar system. Pieces (large and small) of stones and iron (Fe) alloy have been falling from the sky from historical times. They were recognized to be of extraterrestrial origin only in the late nineteenth century. These are called meteorites and are described in some detail in Chapter 8 on the earliest history of the solar system. It is sufficient at this stage to note that the subgroup of meteorites, called the carbonaceous chondrites contain nonbiotic organic compounds, hydrous silicate minerals, and high temperature silicate minerals. They are thought to be the most primitive tangible or condensed solar system matter available for direct analysis in the laboratory.

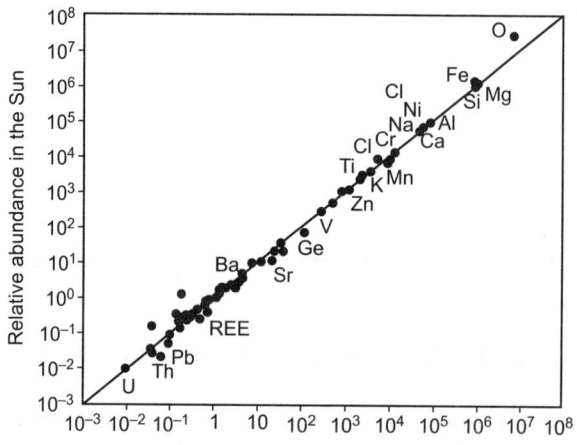

Figure 1.5 Comparison of solar and chondritic abundances with Si set at 10^6 (Data from Anders and Grevessse, 1989)

A comparison of elemental abundances relative to 10^6 atoms of Si between primitive carbonaceous chondrites and solar atmosphere is shown in Figure 1.5 (Anders and Grevesse, 1989). Note the logarithmic scale to cover many orders of magnitude difference in elemental abundances. A perfect agreement would be indicated by a tight alignment of the data to the diagonal line shown. The figure shows a good agreement for most elements, both abundant, like oxygen (O), magnesium (Mg), and Fe, and rare like thulium (Tm) and thorium (Th). The correspondence may be even better than indicated in view of the larger uncertainties in measuring the solar composition by optical spectroscopy. Significant exceptions are the relative depletion of Li and B in the solar photosphere and carbon (C) and N in carbonaceous chondrites, most likely due their volatility. As Li, B, and beryllium (Be) are systematically destroyed in the Sun, chondrites may record the composition of the ancient Sun even better than the present day Sun does. Other important inferences from this figure are:

1. Meteorites, particularly carbonaceous chondrites are extremely valuable links between comochemistry (strictly speaking solar system chemistry) and geochemistry.
2. The meteorites, more than the Sun, show that the pristine solar system material already contained all the 90 elements in the periodic table.
3. Similarity in the relative proportions of H and He in the solar photosphere with other stars indicate little mixing with the helium-enriched and hydrogen-depleted core of the sun (Chapter 3).

Basics

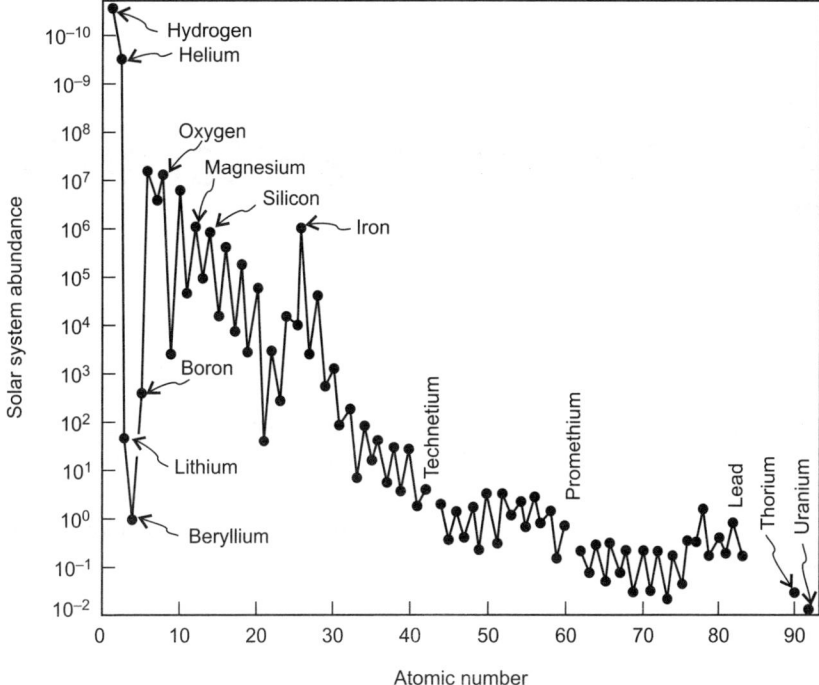

Figure 1.6 Cosmic abundance of elements, relative to 10^6 Si atoms (Anders and Grevesse, 1989).

Anders and Grevesse (1989) have combined the data from spectroscopic analyses of sunlight and chemical analyses of carbonaceous chondrites to come up with a table of elemental abundances whose values are said to be accurate to within 10 per cent or better. These abundances can be considered to represent the primitive solar nebula. Figure 1.6 is a plot of the abundances of the elements in the solar system versus their atomic number. The abundances are expressed as the logarithm to the base 10 of the number of each element relative to 10^6 atoms of Si (corresponding to scale division 6). The significant features of this plot are:

1. Hydrogen and He make up more than 98 per cent of all atoms;
2. Lithium, Be and B are anomalously low;
3. Sharp (exponential) decrease of abundances with atomic number;
4. Absence of technetium and promethium;
5. Even atomic number elements are more abundant than their neighbouring odd atomic number elements, giving rise to the saw tooth profile;
6. Prominent peaks or excesses at Fe and lead (Pb), relative to their immediate neighbours; and
7. Extremely low abundance of elements between Bi and U.

These features are clues to how the elements were formed in the material destined to form the solar system, as covered in Chapter 3.

2 Nuclear Transformations

> The discovery of radioactivity provided an additional (and continuous) heat source for earth's interior and revolutionized studies of earth history by permitting quantitative dating.
>
> C H Langmuir and W Broecker (2012)

2.1 Introduction

Certain nuclides with a specific combination of protons and neutrons may transform spontaneously or be transformed by external agents to another nuclide with a different combination of neutrons and protons. We will consider such transformations as relevant to radiometric dating in the present chapter.

2.2 Spontaneous Nuclear Transformations

2.2.1 Radioactivity

The spontaneous transformation or decay of a potentially unstable nuclide to a more stable nuclide is called radioactivity. The energy released by the decay is carried mainly by particles and radiation. The decaying nuclide and its product nuclide are customarily labelled parent (p) and daughter (d), respectively. If d is radioactive, it decays to another nuclide until a stable d is produced. The nuclear decay process obeys the following conservation laws of Physics: (1) Mass/energy, (2) Electric charge, (3) Linear momentum, (4) Angular momentum, and (5) Nucleon number (Kaplan, 1955; Beiser, 1973; Leighton, 1959).

It was noted in the first chapter that stable nuclides define a narrow band in a Z vs N plot (Figure 1.3) corresponding to the greatest stability of Z/N ratio as a function of N. Unstable nuclei that deviate from the path of stability eventually transform into stable nuclei by different decay modes and rates. A diagonal section across the path of the stability valley will contain isobaric nuclides with the most stable isobar in or close to the path. Neutron-rich isobars will fall below the path and proton-rich isobars above it. Unstable nuclides heavier than ^{209}Bi along the path are also neutron-rich. Although nuclides decay in many modes, the modes most common and relevant to radiogenic isotope geochronometry are few (Dalrymple, 1991), and are shown in Figure 2.1.

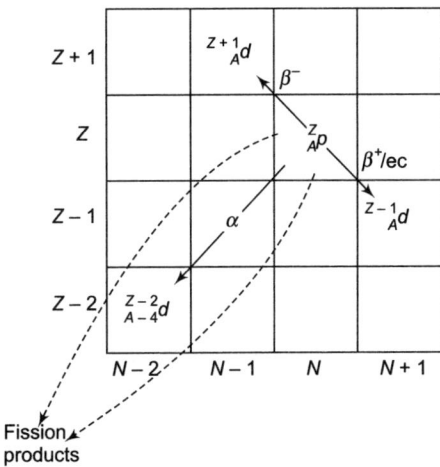

Figure 2.1 The common types of radioactive decay used in radiometric dating and their effect on mass number A, atomic number Z, and neutron number N. Parent isotope p and daughter isotope d.

2.2.2 Beta (β⁻) decay

Neutron-rich nuclei below the stability zone reach stability by converting one of their neutrons into a proton, an electron (β^-), and an antineutrino ($\bar{v}$), and then emitting the latter two. The electron created in the nucleus just before its emission is called a beta or β particle to distinguish it from an orbital or extra-nuclear electron. The Beta (β^-) decay process is described by:

$$p(N, Z) \rightarrow d\,(N-1, Z+1) + \beta^- + \bar{v} + Q \qquad (2.1)$$

Conservation of energy requires that the decay energy (Q) is shared between β and $\bar{v}$. With no electrical charge or mass, the neutrino and its antiparticle, antineutrino, interact very weakly with matter. The beta decay process is mediated by the weak nuclear force.

For example: ^{87}Rb $\rightarrow$ ^{87}Sr + β^- + $\bar{v}$ + Q

If left in an excited state, the daughter nuclide will return to the ground state releasing its excitation energy either in a gamma ray or to an orbital electron in a process known as internal conversion. A gamma ray is simply a high-energy photon (i.e. electromagnetic radiation). Its frequency is related to the energy difference between the upper (excited) and lower (ground) states and Plank's constant ($h/2\pi$).

2.2.3 Beta (β⁺) Decay

Proton-rich nuclei in the upper side of the stability zone approach stability by converting one of their protons into a neutron, positron (β^+) and neutrino (v). β^+ is an antiparticle of an electron, called positron. The beta (β^+) decay process is expressed as:

$$p(N, Z) \rightarrow d(N + 1, Z - 1) + \beta^+ + v + Q \tag{2.2}$$

On coming to rest, the positron interacts with an electron, resulting in the conversion of their total rest mass (mutual annihilation) into two 0.51 MeV gamma rays. This annihilation energy of 1.02 MeV adds to the total energy released by decay.

For example: $^{40}K \rightarrow {}^{40}Ar + \beta^+ + v + Q$

2.2.4 Electron capture decay

An alternative mechanism for a nuclide to decrease its proton number is to capture an orbital electron and emit only a neutrino. An outer orbital electron will fill the vacancy of the captured electron and emit a characteristic x-ray. This x-ray can eject another electron (Auger electron) from the atom. The Electron Capture (EC) process is described as:

$$p(N, Z) \rightarrow d(N + 1, Z - 1) + v + Q$$

For example: $^{40}K \rightarrow {}^{40}Ar + v + Q$ \hfill (2.3)

Both β^+ and EC processes yield the same daughter nuclide. But the former requires decay energy greater than 1.02 MeV.

2.2.5 Branched decay

If an unstable nuclide is flanked on either side by stable isobars, it can decay by branching into both. The branched decay process is described as:

$$p(N, Z) \rightarrow d^1(N + 1, Z - 1) + \beta^+ + v \tag{2.4}$$

$$p(N, Z) \rightarrow d^2(N - 1, Z + 1) + v \tag{2.5}$$

where, d^1 and d^2 are the isotopes of two different elements.

2.2.6 Alpha decay

As the higher mass side of the stability zone closes at $N = 126$ and $A = 209$, heavier nuclei cannot reach the stability zone by emitting β^- or β^+ particles. So, they approach it by emitting an alpha (α) particle (^{4}He), as it reduces Z, N, and A of the parent nuclide by 2, 2, and 4 units, respectively. If left in an excited sate, the daughter will emit a gamma to return to the ground state. The α decay process is:

$$p(N, Z) \rightarrow d(N - 2, Z - 2) + \alpha + Q \tag{2.6}$$

For example: $^{147}Sm \rightarrow {}^{143}Nd + \alpha + Q$

It may have been noted that the above common types of radioactivity result in emission from the parent nucleus of β (negative or positive) particles or an α ($2p + 2n$) particle, but never of just one proton or one neutron.

2.2.7 Spontaneous fission

Very heavy nuclides like ^{238}U decay mainly by α particle emission. But occasionally they split or fission spontaneously in a single step into two product nuclei of roughly equal mass and emit neutrons, gamma rays, and a very large amount of energy, ~200 MeV per fission (Figure 2.1).

The probability of spontaneous fission of ^{238}U is about a million times smaller than that for α decay. In heavier transuranic elements, spontaneous fission is the principal mode of decay. One such nuclide of cosmochronological interest is ^{244}Pu.

2.3 INDUCED NUCLEAR TRANSFORMATIONS

Induced nuclear transformations are classified into Low- and High-energy nuclear reactions.

2.3.1 Low energy nuclear reactions

Induced nuclear reaction refers to the transformation or transmutation of a nuclide to another by bombarding the former with suitable projectiles, usually neutrons, protons, and α particles with energies less than a few MeV. In 1919, Rutherford (1919) demonstrated induced nuclear reaction on nitrogen (N) by bombardment with α particles from a natural radioactive source as:

$$^{14}\text{N} + {}^{4}\text{He} \, (\alpha \text{ particle}) \rightarrow {}^{17}\text{O} + {}^{1}\text{H} \, (\text{proton}) \tag{2.7}$$

This is written concisely as ^{14}N $(\alpha, p)^{17}$O. ^{14}N and ^{17}O of oxygen are target and product nuclei, respectively, α and p incident (projectile) and outgoing (emitted) particles, respectively. Positively charged projectiles like protons and α particles must have sufficient kinetic energy to overcome the electrostatic barrier surrounding the positively charged target nuclide before colliding with it. Neutrons, however, even with very low energies (thermal neutrons) can induce nuclear reactions as they have no electrical charge to be repelled by the positive charge of a target nucleus. A typical neutron-mediated nuclear reaction, written concisely, is ^{14}N $(n, p)^{14}$C. Induced reactions also obey the same conservation laws as the radioactive decay process.

The production rate, P of the product nuclide can be expressed as:

$$P = N_t f \sigma$$

where N_t is the number of target nuclides, σ the probability of occurrence of the reaction known as nuclear cross section in the unit, Barn (= 10^{-24} cm^2), and f is the flux of projectiles. The value of nuclear cross section depends on the projectile type and energy, and target nucleus. If the product nuclide is stable, its concentration will increase linearly with time, t of bombardment, as Pt. If it is unstable (radioactive) its concentration will increase until its production rate is exactly balanced by its own intrinsic decay rate, $N_p \lambda$ (Chapter 5):

$$P = N_p \lambda, \text{ that is, } N_p = P/\lambda$$

where, (P/λ) gives the steady state or equilibrium concentration of the target nuclide. λ is a constant characteristic of the product nuclide (Chapter 5).

2.3.2 Induced nuclear fission

The second type of induced nuclear reaction pertinent to geochronometry is induced fission as distinct from spontaneous fission mentioned earlier. Of the two isotopes of U, only ^{238}U undergoes spontaneous fission, albeit very rarely. The other isotope, ^{235}U does not show any detectable spontaneous fission, but readily undergoes fission when induced (coaxed may be appropriate) by even thermal neutrons. Neutron irradiated ^{235}U atoms split into two nuclides of roughly equal mass. This comes in handy in the Fission Track method of dating minerals and rocks (Chapter 5).

2.3.3 High energy nuclear reactions

If the bombarding particle has very high energy, the target nucleus may be shattered, leaving stable and unstable nuclei of much lower atomic mass and number, as well as protons and neutrons. Unstable particles, such as muons, pions, etc., are also created. Such violent reactions are called spallation. For example, bombardment with protons with energies in the range of billion electron volts (GeV) of ^{56}Fe of the iron nuclei leads to the following reaction (Allegre, 2008):

$$^{56}Fe + p \rightarrow {^{36}Cl} + 3p + 2{^{4}He} + {^{3}He} + 4n$$

The numerous secondary neutrons can induce further reactions. It may be helpful for our purposes to imagine that the target nucleus is just chipped in low energy reactions, split into two fragments of approximately the same mass in induced fission reactions, and shattered into many pieces in high energy (spallation) reactions.

2.4 Induced Nuclear Transformations in the Laboratory and Nature

A copious source of both fast and slow neutrons for laboratory experiments is a nuclear reactor. A sophisticated device called an accelerator is required to endow charged particles like protons and α particles with the energy required to induce nuclear reactions in the laboratory. Many types of accelerators have been developed for nuclear physics experiments, the well known cyclotron being one. Modern accelerators can propel charged particles to GeV energy range.

A natural source of protons and α particles is the cosmic rays, which continuously bombard the earth and other objects in the solar system. Cosmic rays are high energy (several 10^9 up to 10^{19} eV) charged particles, consisting mainly of the nuclei of hydrogen (H) (protons) and He (α particles) with a very minor contribution from heavier elements. A significant fraction of cosmic rays with much lower energies originates from the sun. The origin of the other fraction called galactic cosmic rays is still unclear, but, as the name implies, it is believed to originate from the interstellar medium.

We will now consider a few induced nuclear reactions of geochronometric relevance produced both in the laboratory and in nature:
- $^{14}N(n, p)^{14}C$
- $^{39}K(n, p)^{39}Ar$

These two reactions, when carried out in the laboratory using neutrons from a nuclear reactor illustrate the basic principle of an important technique in geochemical analysis, known as Neutron Activation (Faure and Mensing, 2005). ^{14}C of carbon and ^{39}Ar of argon are radioactive and, hence, can be easily measured by counting their decays. Therefore, they serve as proxies for the direct measurement of nitrogen or potassium in a sample of interest. Nitrogen and potassium are not usually measured this way, but partial conversion of one isotope of potassium into an isotope of Ar by fast neutron irradiation in a nuclear reactor is the basis of an elegant refinement of one of the common radioactive dating methods, as we will see in chapters 5 and 9.

These and many other reactions take place naturally by the interaction of galactic cosmic rays with the atmosphere and top silicate layers of the earth, respectively. When galactic cosmic rays strike the top of the atmosphere, they cause spallation or shattering of target nuclides therein, releasing cascades of secondary particles including neutrons, protons, and mu–mesons. These secondary particles cause additional reactions in the lower atmosphere resulting in the production of stable and unstable isotopes of many elements. A small fraction of the secondary radiation actually reaches the surface of the earth to produce new nuclides in the exposed soil and rock (Lal, 1988). We will see in Chapter 3 that all the isotopes of chemical elements in the solar system are ultimately cosmogenic in origin in the wider sense that they were produced in cosmic settings.

2.5 ROLE OF NATURAL RADIOACTIVITY IN GEODYNAMICS AND GEOCHRONOLOGY

Radioactive decay can be described in a general way as:

$$\text{Radioisotope} \rightarrow \text{Radiogenic isotope} + \text{Radiation} + \text{Energy}$$
$$(p) \qquad\qquad (d) \qquad\qquad (\alpha, \beta, \gamma) \qquad Q \qquad (2.8)$$

The above equation implies that as a radionuclide decays within a solid medium, the energy carried by the radiation is dissipated as heat within the medium, and the product nuclide accumulates close to its parent. This means that decay of radionuclides within the earth results in heat generation and accumulation of stable products with time. The first is the basis of the dynamism of the earth, and the second is the basis of geochronometry. We will briefly consider radioactive heat production within the earth, and defer the geochronometric applications of natural radionuclides to Chapter 5.

2.5.1 Heat generation

Heat generation in the earth is mainly by the decay of the long-lived isotopes, ^{238}U, ^{235}U, ^{232}Th and ^{40}K, whose natural abundances are given in Table 2.1.

Table 2.1 Abundances of the Radioactive Isotopes of Uranium, Thorium and Potassium

Isotope	Abundance (%)	Heat production (µW/Kg)
^{238}U	99.28	93.63
^{235}U	0.72	568.50
^{232}Th	100.00	26.92
^{40}K	0.016	27.92

Heat production/Kg in the earth, assuming its U, Th, and K concentrations are as in chondritic meteorites (Chapter 1), namely, U = 0.015 ppm, Th = 0.045 ppm, and K = 900 ppm, is:

$$93.63 \times 0.9928 \times 0.015 \times 10^{-6} + 568.5 \times 0.0072 \times 0.015 \times 10^{-6}$$
$$+ 26.92 \times 0.045 \times 10^{-6} + 27.92 \times 0.016 \times 10^{-2} \times 900 \times 10^{-6} = 5.58 \text{ µW/Kg}$$

In the entire earth, $5.58 \times 10^{-6} \times 5.97 \times 10^{24} = 33.3 \times 10^{12}$ W = 33.3 TW or 1.1×10^{21} J/yr.

The measured surface heat flow is ~41 TW. Thus, radiogenic heat represents a substantial component of the earth's internal heat budget, which drives all geodynamic processes such as plate tectonics, volcanism, and earthquakes.

2.6 Statistical Aspect of Radioactivity

The occurrence of nuclear decays is a random phenomenon. This can be seen from the fluctuations in the number of decays recorded in successive uniform time intervals in a radioactive source. Let n successive counts per minute be $x_1, x_2 \ldots x_n$. The average, or arithmetic mean of these counts is, by definition,

$$\bar{x} = \frac{1}{n}(x_1 + x_2 + \ldots x_n) = \left(\frac{1}{n}\right)\Sigma x_i \qquad (2.9)$$

The symbol Σ stands for summation of terms from $i = 1$ to $i = n$. The mean value does not indicate the extent of scatter among the values. The deviation of any value from the mean is:

$$\Delta_i = (x_i - \bar{x}),$$

The mean or average of the n individual deviations is not a measure of the spread in the data because it is just zero. The mean of the squares of the individual deviations is called the variance, and is given by:

$$\sigma_x^2 = \left(\frac{1}{n}\right)\Sigma\Delta_i^2 = \left(\frac{1}{n}\right)\Sigma(x_i - \bar{x})^2 \qquad (2.10)$$

The square root of the variance, σ_x is called the standard deviation of the data set, and is usually used as a measure of the data dispersion.

Before we consider the statistics of counting data, we must introduce the concept of probability. Probability is defined in many ways. If an outcome x_i can happen in n_i ways out of a total n possible equally likely ways or outcomes, the probability of occurrence of outcome, x_i (called success) is denoted by:

$$p(x_i) = \left(\frac{n_i}{n}\right)$$

For example: Let e be the event that the numbers 3 or 4 turn up in a single toss of a die. There are six ways in which the die can fall, resulting in the numbers 1, 2, 3, 4, 5, or 6. If the die is not loaded, these six ways are equally likely. Since e can occur in two of these ways, we have: $p = \text{Pr}(e) = 2/6 = 1/3$.

The probability of nonoccurrence of outcome x_i (called its failure) is denoted by:

$$q(x_i) = \frac{(n - n_i)}{n} = 1 - p(x_i)$$

The probability of an outcome is, therefore, a number between 0 and 1. If the outcome does not happen, its probability is 0. If it must happen, that is, its occurrence is certain, its probability is 1. According to this definition, the relative frequency of occurrence of an outcome may also be considered its probability, when n increases indefinitely. For this reason we can think of probability distribution as being distributions of 'populations', whereas, relative frequency distributions are distributions of 'samples' drawn from the population.

Nuclear Transformations

Reverting to nuclear counting data, if the values $x_1, x_2, \ldots x_i$ occur $n_1, n_2, \ldots n_i$ times, respectively, in a large number of observations, their mean can be rewritten in terms of their relative frequencies as:

$$\bar{x} = \left(\frac{n_1}{n}\right) x_1 + \left(\frac{n_2}{n}\right) x_2 + \cdots \left(\frac{n_i}{n}\right) x_i \tag{2.11}$$

$$= p(x_1) x_1 + p(x_2) x_2 + \cdots p(x_i) x_i \tag{2.12}$$

$$= \Sigma p(x_i) x_i \tag{2.13}$$

in the limit of very large n. Similarly, the definition of variance is:

$$\sigma_x^2 = \Sigma p(x_i)(x_i - \bar{x})^2 \tag{2.14}$$

where, $p(x_i)$ is called a probability distribution function and gives the fraction of observations corresponding to an outcome, x_i. This may be generalized to give the average value (the expectation value) of any function $f(x)$ as:

$$f(\bar{x}) = \Sigma f(x_i) x_i \tag{2.15}$$

If $p(x)$ is a continuous function of x, the mean and standard deviation become:

$$\mu = \int p(x) \, x dx \tag{2.16}$$

$$\sigma^2 = \int p(x)(x - \mu)^2 dx \tag{2.17}$$

Two or more events, e_1 and e_2 are called mutually exclusive, if the occurrence of any one of them excludes the occurrence of the others. In general, if $e_1, e_2, \ldots e_n$ are n mutually exclusive events having respective probabilities of occurrence, $p_1, p_2, \ldots p_n$, then the probability of occurrence of either e_1 or e_2 or $e_3 \ldots$ or e_n is $p_1 + p_2 + p_3 + p_4 + \cdots p_n$.

For example: When a coin is tossed, the probability of either head or tail is ½ + ½ = 1.

Two or more events are independent if the occurrence or nonoccurrence of e_1 does not affect the probability of occurrence of e_2. In general, if $e_1, e_2, \ldots e_n$ are n independent events having respective probabilities $p_1, p_2, \ldots p_n$, then the probability of occurrence of $e_1, e_2, e_3, \ldots e_n$ is $p_1, p_2, p_3, \ldots p_n$.

For example: If a coin is tossed twice, the probability of getting heads twice is $(1/2) \times (1/2) = (1/4)$.

2.6.1 Binomial distribution

The binomial distribution law treats a fairly general case of compounding of probabilities (Friedlander and Kennedy, 1955). If p is the probability that an event will happen in any single trial (called the probability of a success) and $q = 1 - p$ is the probability that it will fail to happen in any single trial (called the probability of a failure), then the probability that the event will happen exactly r times in n trials [i.e. r successes and $(n - r)$ failures will occur] is given by:

$$p(r) = \left[\frac{n!}{(n-r)! \, r!}\right] p^r q^{n-r} \tag{2.18}$$

where, $r = 0, 1, 2, \ldots n$, $n! = n(n-1)(n-2)\ldots,1,0$ and $0! = 1$. The probability is specified by the two parameters, n and p.

For example: The probability of getting exactly two heads in six tosses of a coin is:

$$p(2 \text{ heads}) = \left(\frac{6!}{4!2!}\right)\left(\frac{1}{2}\right)^4\left(\frac{1}{2}\right)^{6-4} = \left(\frac{15}{64}\right)$$

For example: The probability of getting at least four heads in six tosses of a coin is:

$$p = \left(\frac{6!}{4!2!}\right)\left(\frac{1}{2}\right)^4\left(\frac{1}{2}\right)^{6-4} + \left(\frac{6!}{1!5!}\right)\left(\frac{1}{2}\right)^5\left(\frac{1}{2}\right)^{6-5} + \left(\frac{6!}{0!6!}\right)\left(\frac{1}{2}\right)^6\left(\frac{1}{2}\right)^0$$

$$= \left(\frac{15}{64}\right) + \left(\frac{6}{64}\right) + \left(\frac{1}{64}\right) = \left(\frac{11}{32}\right)$$

The name for binomial distribution comes from the fact that coefficients correspond to the successive terms in the binomial formula or binomial expansion:

$$(p+q)^n = p^n + np^{n-1}q + \left[\frac{n(n-1)}{2!}\right]p^{n-2}q^2 + \ldots + q^n$$

The average value to be expected is obtained from Equation (2.13) as:

$$\bar{r} = \Sigma p(r)r = \Sigma r\left[\frac{n!}{(n-r)!r!}\right]p^x q^{n-r} \qquad (2.19)$$

To evaluate this summation consider the binomial expansion of $(px + q)^n$:

$$(px+q)^n = \Sigma\left[\frac{n!}{(n-r)!r!}\right]p^r x^r q^{n-r} = \Sigma x^r p(r) \qquad (2.20)$$

Differentiating with respect to x, we get:

$$np(px+q)^{n-1} = \Sigma rx^{r-1}p(r) \qquad (2.21)$$

Letting x = 1 and using q = (1 − p), we get the desired expression:

$$np = \Sigma rp(r) = \bar{r} \qquad (2.22)$$

We could have guessed this result. If the probability A coming in a single event is p, we would expect the average number of As to be (np). To get the standard deviation σ_r for this expected average value, $\bar{r}$, we differentiate Equation 2.21 again, with respect to x:

$$n(n-1)p^2(px+q)^{n-2} = \Sigma r(r-1)x^{r-2}p(r)$$

Again, letting x = 1 and using p + q = 1, we have:

$$n(n-1)p^2 = \Sigma r(r-1)p(r) = \Sigma r^2 p(r) - \Sigma rp(r)$$

$$n(n-1)p^2 = \overline{r^2} - \bar{r}$$

Because variance σ^2 is given by:

$$\sigma_r^2 = \overline{r^2} - \bar{r}^2$$

$$\sigma_r^2 = n(n-1)p^2 + \bar{r} - \bar{r}^2$$

Nuclear Transformations

With $\bar{r} = np$, we finally get:

$$\sigma_r^2 = n^2p^2 - np^2 + np - n^2p^2 = np(1-p) = npq$$

$$\sigma_r = \sqrt{npq} \qquad (2.23)$$

2.7 Binomial Distribution for Radioactive Disintegrations

If p is the probability for a radioactive atom to decay in time t and $q = (1 - p)$ is the probability for it to have survived time t, then the probability p(x) for exactly x decays out of n radioactive atoms will be given by the binomial distribution as:

$$p(x) = \left[\frac{n!}{(n-x)!\,x!}\right] p^x q^{n-x} \qquad (2.24)$$

The expected average number of atoms decaying in time t will be:

$$\bar{x} = np = n(1 - \exp(-\lambda t)) = n\lambda t \text{ for } \lambda t \ll \text{half-life.} \qquad (2.25)$$

The expected standard deviation,

$$\sigma_x = \sqrt{npq} = [n(1-e^{-\lambda t})(e^{-\lambda t})]^{0.5} = [\bar{x}e^{-\lambda t}]^{0.5} = \sqrt{\bar{x}} \text{ for } \lambda t \ll \text{half-life.}$$

When any individual value is large enough it can be approximated to the average count in order to calculate the standard deviation.

σ for ~100 counts is $\sqrt{100} = 10$.

If counts are recorded for 10 minutes, $\bar{x} \approx 1000$

$$\sigma = \sqrt{1000} = 32.$$

The error in the counting rate is now reduced as:

$$\frac{(1000 \pm 32)}{10} = 100 \pm 3.2 \text{ cpm}$$

Thus, for a given counting rate, the σ for the rate is inversely proportional to the square root of the time of measurement.

2.8 Poisson Distribution

The binomial distribution for radioactive decays asymptotically approaches another well known probability distribution called the Poisson Distribution in the limiting case when n tends to infinity and p tends to zero in such a way that the average np is a constant, a. Neither n nor p is known, but only their product, namely the expected average. The probability p(r) that event A occurs r times is:

$$p(r) = \text{Limit} \left[\frac{n!}{(n-r)!\,r!}\right] p^r q^{n-r} \qquad (2.26)$$

$$p(r) = \left(\frac{a^r}{r!}\right) \text{limit} \left[\frac{n(n-1)(n-2)\ldots(n-r+1)}{n^r}\right] q^{n-r}$$

As n tends to infinity, the term in the first bracket tends to unity, as r is finite and small compared to n.

Also, $\quad q^{n-r} = (1-p)^{n-r} = \left[1 - \left(\dfrac{a}{n}\right)\right]^{n-r}$

$\qquad\qquad = e^{-a}$ as n tends to infinity:

$$p(r) = \frac{e^{-a} a^r}{r!}, \qquad (2.27)$$

which is the Poisson Distribution. In words, the probability of obtaining a particular number of counts is given by:

$$p(r) = \frac{e^{-a} a^r}{r!}$$

where, a is the average rate to be expected. Note that, whereas, the binomial distribution is specified by two parameters, the Poisson Distribution is specified by a single parameter, a.

Mean and standard deviation

$$\bar{r} = Np = a \qquad (2.28)$$

$$\sigma = \sqrt{Npq} = \sqrt{a}, \qquad (2.29)$$

since q = 1 in the limit. As a becomes large, the Poisson Distribution becomes more and more symmetric about a. The Poisson Distribution applies to the counting of particles when the average is constant, as in the case of nuclear and atomic physics.

3 Nucleosynthesis

> I think there should be a law of Nature to prevent a star from behaving in this absurd way.
>
> **Arthur Eddington**

3.1 Introduction

The variety and relative abundances of the 90 elements in the Solar System must have been produced somewhere else before its formation about 4.6 by ago. The earliest time some or all of them could have been produced somewhere is 13.7 billion of years ago when the universe is believed to have originated in a spectacular explosion—Big Bang—of what is known as a singularity in cosmological parlance. At least three sets of observations support the Big Bang theory.

1. Expanding universe Hubble discovered that galaxies are receding from each other at a speed proportional to the distance between them, based on red shifts of light coming from distant galaxies. This implies that at some time in the past all the matter in the universe must have been concentrated at one point.

2. Big Bang nucleosynthesis According to the Standard Cosmological model, the matter in the universe 30 minutes after the explosion must have consisted mostly of hydrogen (H) and helium (He) with only traces of deuterium (^{2}H) and helium (^{3}He), as confirmed by recent observations.

3. Cosmic microwave background Theory also predicts that the universe must now be filled with an isotropic background radiation in the microwave region of the electromagnetic spectrum as a relic of the initial radiation from the Big Bang. Penzias and Wilson (Penzias and Wilson, 1965) have indeed detected such a cosmic microwave background radiation corresponding to black body emission at 2.7 K.

The predicted sequence of events within the first 3 minutes of the Big Bang, according to the Standard Cosmological model, is given in Table 3.1.

Table 3.1 Predicted Sequence of Events Within the First Three Minutes of the Big Bang

Time	Temperature, K	Event
0	------	Singularity before explosion
10^{-50} s	------	Extremely rapid initial explosion called Inflation. The four fundamental forces are unified.
10^{-43} s	10^{32}	Gravitational force separates
10^{-35} s	10^{28}	Strong force separates
10^{-10} s	10^{15}	Weak and electromagnetic forces separate
10^{-4} s	10^{12}	Proton and neutron abundances reduced
3 m	10^9	Formation of helium nuclei
5×10^5 y	3×10^3	Formation of H and He atoms and the first batch of galaxies and stars
4.5×10^9 y	2.7	Huge number of galaxies and stars bathed in the relic of Big Bang radiation

3.2 STELLAR NUCLEOSYNTHESIS

For geologists used to events on a time scale of at least a few thousand years, incredibly short time intervals like 10^{-50}, 10^{-43}, and 10^{-35} s make no sense. Even a comparatively much longer time interval like 10^{-10} s is so short that light with a velocity of as high as 3,00,000 kilometres per second will cover just about 3 centimetres in this interval. The important point for geologists to note is that Big Bang, or cosmological nucleosynthesis, was restricted to only H and He and to within the first 3 minutes. So the variety and abundance of nuclear species heavier than He in the pristine Solar System material must have been produced somehow, somewhere between 13.7 and 4.6 billion years ago. That stars are the most likely sites for their synthesis and later widespread dispersal comes from the following crucial astronomical observations:

1. The source of prodigious energy radiated by stars, like the Sun, must be nuclear.
2. All stars, regardless of their age, have about the same He/H ratio as produced in the cosmological or primordial nucleosynthesis.
3. Very old stars are poor in heavy elements.
4. Stars have variable composition, possibly reflecting a gradual build-up of heavy elements in the interstellar medium.
5. The elements technetium (Tc) and promethium (Pm) are absent in the Sun and meteorites, as all their isotopes are short-lived ($<10^6$ y). So the presence of these elements in some massive stars and nebulae, as revealed in their optical spectrum, reflects their active synthesis in them.
6. There is an explosive end of massive stars that disperses synthesized elements into the interstellar medium, possibly to be recycled into later stars.

Nuclear physics considerations bearing on possible nucleosynthetic pathways are:
1. Formation of multi-proton nuclides requires that protons are first brought sufficiently close together against their mutual electrostatic or Coulomb repulsion before the short-range attractive nuclear force can fuse them together (Chapter 1). A clue to this step comes from

nuclear reactions in the laboratory induced by bombarding target nuclei with energetic charged particles like protons and alpha (α) particles (Chapter 2). In contrast, electrically-neutral neutrons need not be so energetic to induce nuclear transformations. Kinetic theory shows that proton-proton collisions are possible at a temperature of about 10^7 K.

2. The binding energy per nucleon versus mass number curve (Figure 1.4) shows a steep rise for light nuclei up to iron (Fe) and a gradual and smooth decrease for nuclei heavier than Fe. This implies that fusion of light nuclei to form heavier nuclei up to Fe is energetically favoured, while fusion of charged particles with heavier nuclei is not so.

3. The type and rates of nuclear reactions that are invoked to account for generation of energy and elemental abundances in stars in different stages of their evolution must be consistent with experimental data obtained in the laboratory.

Stars of various masses relative to the sun are formed when an interstellar cloud of gas and dust contracts gravitationally, due either to random local density inhomogeneities in the cloud, or shock waves spreading from nearby stellar explosions (supernovae). The key features of stars from an astronomical angle are their mass, luminosity, surface temperature, and distance from us. These observational parameters are used to classify stars and place them into an evolutionary sequence. A widely used classification diagram is the Hertsprung-Russell diagram, or simply the H-R diagram which is a plot of stellar luminosity versus effective temperature. The luminosity of a star is a function of its radius and effective temperature. The surface temperature is determined from its colour, and is extremely different from its interior temperature. For example, the Sun's surface temperature is about 5,700 K, but its core temperature is 14×10^6 K. A simplified H-R diagram is shown in Figure 3.1 (Shu, 1982). Most stars fall within a narrow band called the Main Sequence.

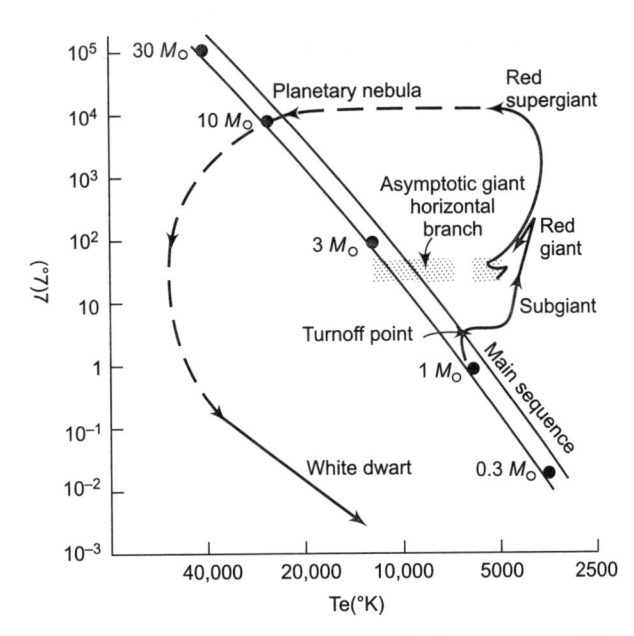

Figure 3.1 Hertzsprung-Russell diagram of luminosity, L vs effective temperature, Te in degrees Kelvin (K). The solid line shows the evolution of a low mass star, like the Sun, from the main sequence to white dwarf. The dashed line indicates the area of uncertainty in the path from the asymptotic giant to the white dwarf (Shu, 1982).

Stars are born and spend most of their lifetimes on the main sequence. The precise evolutionary path of a star from the main sequence depends mainly on its mass. Figure 3.1 shows the probable evolution of a low mass star like the Sun (1 M_o).

Present-day models of stellar nucleosynthesis are based on a classic review paper by Burbridge et al. (1957), in which eight different element-building processes were linked to different stages of stellar evolution. These are briefly described below.

3.2.1 Hydrogen burning

The release of potential energy due to gravitational contraction heats up the core of a protostar. When the core temperature reaches 10^7 K, protons will be sufficiently energetic to overcome their mutual repulsion and fuse in a few steps to He, resulting in the release of energy to light up the star. This process, called hydrogen burning (in the nuclear parlance), consists of the three steps:

$$^1H + {}^1H \rightarrow {}^2H + \beta^+ + \nu + Q \tag{3.1}$$

$$^2H + {}^1H \rightarrow {}^3He + \nu + Q \tag{3.2}$$

$$^3He + {}^3He \rightarrow {}^4He + 2\,{}^1H + \nu + Q \tag{3.3}$$

where, Q is the energy output of each stage. The first two steps must occur twice before the last can take place. Deuterium (H-2) and He-3 generated in the first two steps are used up in the third. Total energy released per fusion is ~23 MeV. The core of the star stabilizes in size because of the balance between energy production in the core and its loss as radiation from the surface (Dickin, 2005). A star spends most of its life in this phase, the Main Sequence (Figure 3.1). Our Sun will probably last ~10^{10} y.

If a star inherits traces of carbon (C) from the debris of nuclear 'ash' of ancestral stars, a more efficient and faster hydrogen burning process, called the Carbon-Nitrogen-Oxygen (CNO) cycle is possible.

$$^{13}C + {}^1H \rightarrow {}^{13}N + \nu \tag{3.4}$$

$$^{13}N \rightarrow {}^{13}C + \beta^+ + \nu \text{ (beta decay)} \tag{3.5}$$

$$^{13}C + {}^1H \rightarrow {}^{14}N + \nu \tag{3.6}$$

$$^{14}N + {}^1H \rightarrow {}^{15}O + \nu$$

$$^{15}O \rightarrow {}^{15}N + \beta^+ + \nu \text{ (beta decay)} \tag{3.7}$$

$$^{15}N + {}^1H \rightarrow {}^{12}C + {}^4He \tag{3.8}$$

The second and fifth steps are beta decays of short-lived ^{13}N and ^{15}O. Each step in the chain needs to occur only once for the hydrogen burning fusion. As ^{12}C is conserved, it acts only as a catalyst. The CNO cycle requires a higher temperature than the above p-p chain because of the higher electrical charges of C and nitrogen (N) relative to a proton. The CNO cycle contributes less than 10 per cent hydrogen burning in a star of solar mass, but dominates in more massive stars accounting for their shorter residence in the Main Sequence.

3.2.2 Helium burning

When the H in the core is exhausted, energy production in the core will decrease and so will the outward radiation pressure. The core will begin to contract under gravitational force. The resulting rise in core temperature causes the still H-rich outer layers of the star to expand. The large increase in the surface area of the star means that energy will be radiated at a lower surface temperature, ~4,000 K (as against 6000 K during the Main Sequence). The star appears large and redder and, hence, is called a 'Red Giant'. When the core temperature rises to 10^8 K at some point during the Red Giant stage, the He nuclei in the core will become energetic enough to fuse into ^{8}Be, as:

$$^4\text{He} + ^4\text{He} \leftrightarrow ^8\text{Be} \tag{3.9}$$

But as ^{8}Be has an extremely short life (~10^{-16}s), it will disintegrate almost as soon as it is created. This is the reason for the cosmological (Big Bang) synthesis to be blocked at ^{4}He. However, as there are about 10^9 ^{4}He nuclei for every ^{8}Be nuclide in the core, a third He nuclide can collide with the ^{8}Be nuclide before it decays, leading to the formation of a ^{12}C nuclide according to:

$$^8\text{Be} + ^4\text{He} \rightarrow ^{12}\text{C}* \tag{3.10}$$

$$^{12}\text{C}* \rightarrow ^{12}\text{C} + \gamma + Q \tag{3.11}$$

where, ^{12}C* represents ^{12}C nuclide in an excited state. It is the gamma ray emitted from the de-excitation of ^{12}C* that sustains the core temperature. This near-simultaneous fusion of three He nuclei is aptly called the Triple alpha process. An exactly simultaneous collision of three α particles to form ^{12}C, skipping the intermediate ^{8}Be, is extremely improbable despite the high density of the He nuclei. The net energy, Q released by this fusion is 7.65 MeV, about four times smaller than that from hydrogen burning. However, if it were not for the Triple alpha process the synthesis of elements heavier than He would have been impossible.

Following the synthesis of ^{12}C, further helium burning reactions proceed as follows:

$$^{12}\text{C} + ^4\text{He} \rightarrow ^{16}\text{O} + \nu \tag{3.12}$$

$$^{16}\text{O} + ^4\text{He} \rightarrow ^{20}\text{Ne} + \nu \tag{3.13}$$

$$^{20}\text{Ne} + ^4\text{He} \rightarrow ^{24}\text{Mg} + \nu \tag{3.14}$$

The elements lithium (Li), beryllium (Be), and boron (B) are rare in the hot stellar interiors, because they quickly combine with protons to form two He nuclei, for example:

$$^7\text{Li} + ^1\text{H} \rightarrow 2\ ^4\text{He} \tag{3.15}$$

These elements are, therefore, bypassed by stellar nucleosynthetic reactions. Their extremely small abundance in the Solar System material is believed to be due to the fragmentation of heavy elements in the galactic cosmic rays (Chapter 2) as they hit interstellar atoms.

The helium burning stage lasts about a tenth as long as the Main sequence stage. When He is exhausted, energy production drops. Small stars like the Sun quickly lose their outer layers to space and fade away as White Dwarfs (Figure 3.1).

3.2.3 Carbon and oxygen burning

After He is exhausted in the core of a Red Giant star of greater than 3 solar masses, the core contracts and temperature rises sufficiently for ^{12}C to fuse and form ^{24}Mg of magnesium, ^{23}Na of sodium, and ^{20}Ne of neon, as:

$$^{12}\text{C} + ^{12}\text{C} \rightarrow ^{24}\text{Mg} + \nu \tag{3.16}$$

$$^{12}\text{C} + ^{12}\text{C} \rightarrow ^{23}\text{Na} + ^1\text{H} \tag{3.17}$$

$$^{12}\text{C} + ^{12}\text{C} \rightarrow ^{20}\text{Ne} + ^4\text{He} \tag{3.18}$$

The H and He nuclei so generated leads to further reactions to fill in gaps between masses 12 and 24. In more massive stars (> 8 solar mass) the temperature in the core rises to 10^9 K, igniting oxygen burning:

$$^{16}O + {}^{16}O \rightarrow {}^{32}S + \nu \tag{3.19}$$

$$^{16}O + {}^{16}O \rightarrow {}^{31}P + {}^{1}H \tag{3.20}$$

$$^{16}O + {}^{16}O \rightarrow {}^{28}Si + {}^{4}He \tag{3.21}$$

$$^{16}O + {}^{16}O \rightarrow {}^{24}Mg + 2\,{}^{4}He \tag{3.22}$$

The α process can build-up from ^{24}Mg through the sequence ^{28}Si, ^{32}S, ^{36}Ar, and ^{40}Ca (Figure 3.2) where it terminates by the instability of ^{44}Ti.

3.2.4 Silicon burning or e-process

At a core temperature of ~3×10^9 K, before fusion of two ^{28}Si nuclides can begin to form ^{56}Fe directly, collisions between the already existing nuclei break them up, but the fragments combine with other nuclei aided by protons, α particles, and a high density of high energy photons. These reactions quickly result in an energetic equilibrium which favours a net build-up of Fe group (Fe, Ni, and Co) elements corresponding to the peak of the binding energy curve (Figure 1.4). This equilibrium, or e-process, is very brief, and marks the end of energy generation by fusion.

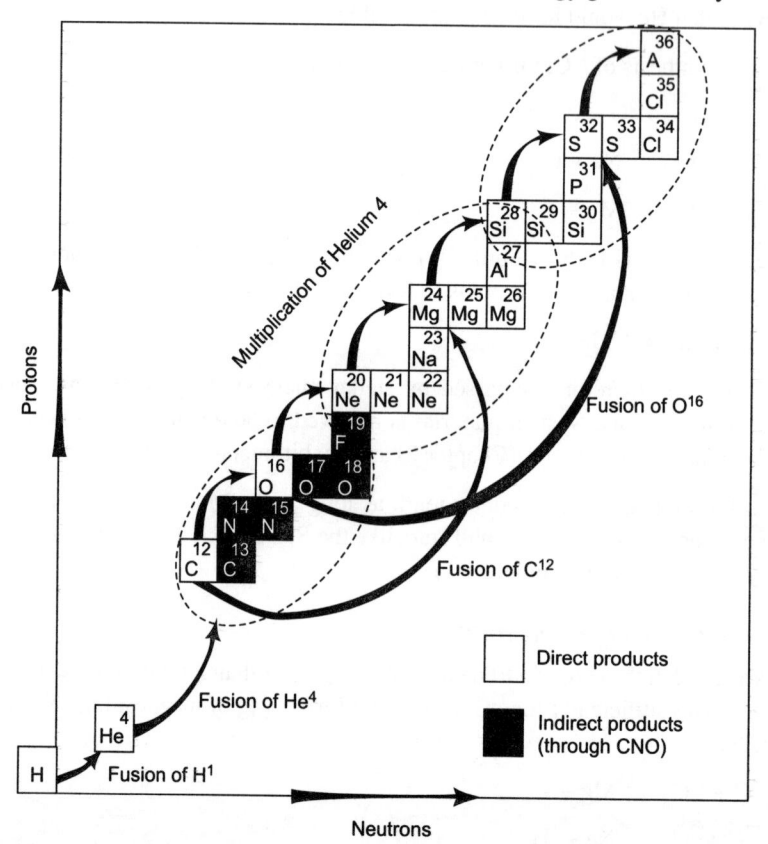

Figure 3.2 An illustration of He, C, and O fusion on the protons vs neutrons diagram (Allegre, 1992).

At the very end of its life, a massive star will have a layered structure with an iron core surrounded by concentric shells of progressively lighter elements (Figure 3.3). Exhaustion of nuclear fuel in the iron core triggers its collapse. The core implodes in a few seconds compressing protons and electrons into neutrons to end up as a small neutron star. The outer layers of lighter elements will first collapse on the imploded core, and then bounce back as shock wave-triggering rapid nuclear reactions in the outer layers. The surge of energy literally blows up the star as a supernova dispersing the debris into interstellar medium.

3.2.5 Production of elements heavier than iron

At the end of formation of Fe group elements a massive Red Giant star would contain only about 20 elements. These 20 elements (excluding the primordial H and He) represent the bulk of the Solar System material. At this stage more than 60 odd elements heavier than Fe that occur in the Solar System are yet to be produced. Many of these have several stable and a few long-lived radioisotopes and are crucial to elemental and isotope geochemistry. Charged particle (protons and alpha particles) reactions to form these elements are endothermic (energy consumers) and, hence, not sustainable, unlike the fusion reactions prior to Fe synthesis. Successive neutron captures by seeds of abundant nuclei have been invoked to build these elements. The time interval between successive neutron additions or captures affects the sequence of heavy elements formation. Neutron additions on a slow time scale are called s-process and on a rapid time scale, r-process.

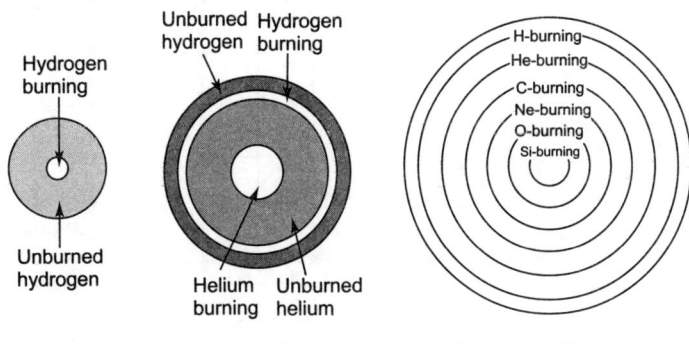

Name of process	Fuel	Products	Temperature
Hydrogen-burining	H	He	60×10^6 °K
Helium-burning	He	C, O	200×10^6 °K
Carbon-burning	C	O, Ne, Na, Mg	800×10^6 °K
Neon-burning	Ne	O, Mg	1500×10^6 °K
Oxygen-burning	O	Mg to S	2000×10^6 °K
Silicon-burning	Mg to S	Elements near Fe	3000×10^6 °K

Figure 3.3 Stars with progressively hotter nuclear fires. The star at the left burns H, the star in the middle burns He to form C and O in its core, the star on the right has multilayered fire all the way up to Si burning to form ^{56}Fe in the core. Approximate temperatures required to ignite the successive fuels are also given (Langmuir and Broecker, 2012).

3.2.6 s-process

During the last few million years before the implosion of the Fe core of a Red Giant star, neutrons are produced in a slow and steady fashion from reactions like:

$$^{13}C + {}^4He \rightarrow {}^{16}O + n + v \tag{3.23}$$

$$^{21}Ne + {}^4He \rightarrow {}^{24}Mg + n + \nu \qquad (3.24)$$

These are captured by nuclei beyond the peak of the Fe group with time intervals between successive captures of 1 to 10^5 years. For example, the first capture will shift ^{56}Fe to its isotope ^{57}Fe, the second capture to ^{58}Fe, and a third capture to ^{59}Fe. But ^{59}Fe has too many excess neutrons to be stable. It will beta (β) decay to isobar ^{59}Co of the next higher element, cobalt. It is now the turn of ^{59}Co to capture a neutron to shift to ^{60}Co, which is unstable, and β decays to the isobar ^{60}Ni. Figure 3.4 shows a portion of the nuclide chart where the s-process trajectory will enter it from the left at ^{84}Kr. The trajectory moves in small zig-zag steps through the stable isotopes ^{85}Rb, ^{86}Sr, ^{87}Sr and ^{88}Sr, ^{89}Y, etc. It is clear from this figure that the s-process path will skip the stable, or nearly stable*s* isotopes ^{82}Se, ^{86}Kr, ^{87}Rb, and ^{96}Zr. The s-process path will climb the path of the greatest stability of proton/neutron ratio, and will finally be terminated by the α decay of ^{210}Po to ^{206}Pb, and ^{209}Bi to ^{205}Tl.

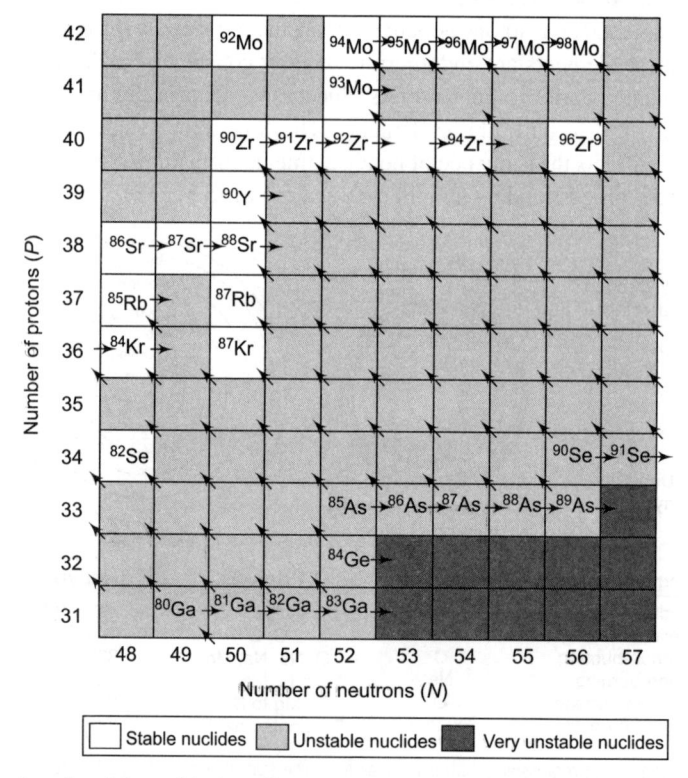

Figure 3.4 A portion of the nuclide chart, illustrating how the s-process (upper set of arrows) and the r-process (lower set of arrows) produce new elements (nuclides) by neutron capture and subsequent decay (Richardson and McSween, 1989).

3.2.7 r-process

In contrast to the step-by-step s-process operating in a quiescent way for millions of years even perhaps in a small star, the r-process takes place in extremely short bursts (a few seconds at most) of colossal fluxes of neutrons during the supernova explosion of a massive Red Giant star. The situation now is not just a single neutron addition to the nucleus, which will either accommodate it or get rid of it by β decay, leading to the next higher element. In the r-process many neutrons are pumped very quickly into a nuclide near the Fe group of elements until it cannot absorb any more. Only then does this neutron-bloated (loaded) nuclide decay promptly to the next higher element. Caught in the still prevailing hail of neutrons, this newly minted nuclide is driven relentlessly to the right on a path quite remote from the path of stability that is tracked by the s-process nuclides. A segment of the r-process path is shown by the lower set of arrows in Figure 3.4. The path enters the chart from the bottom at the unstable neutron-loaded nuclide ^{80}Ga formed by the β decay of ^{89}Zn saturated with 10 excess neutrons relative to the heaviest stable Zn isotope, ^{80}Zn. Four quick neutron additions shift ^{80}Ga of gallium to saturation at ^{84}Ga. ^{84}Ga decays promptly to ^{84}Ge of germanium, which, in turn, decays promptly to ^{86}As of arsenic after a single neutron addition. It is now the turn of ^{86}As to shift quickly to ^{90}As ending with the β decay to ^{90}Se. The r-process path leaves the chart segment at ^{91}Se and zooms past Bi (the end of the s-process path) and even past uranium (U) and thorium (Th), stopping finally only by neutron-induced fission at mass 254, as shown in Figure 3.5. The intermediate mass fragments produced from such fissions are caught in the still raging rain of neutrons and move up the neutron saturation route.

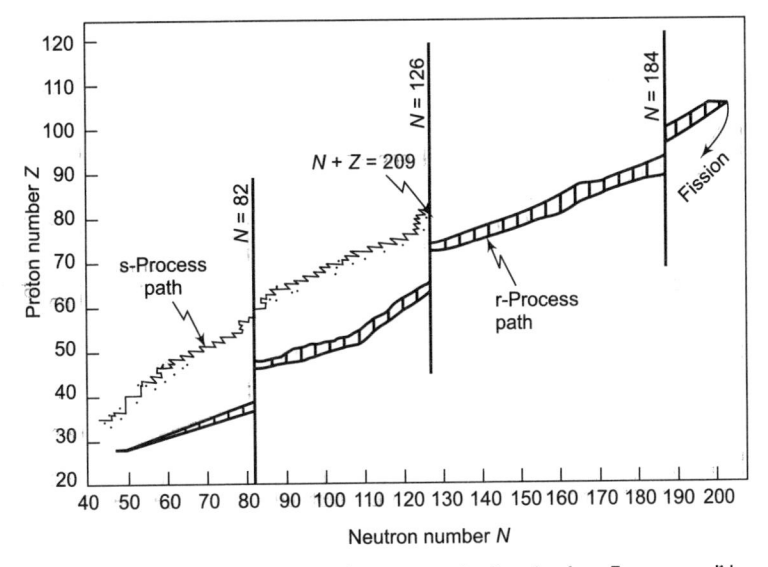

Figure 3.5 The quite different s- and r-process paths diverging from Fe group nuclides (Langmuir and Broecker, 2012).

When the colossal flux of neutrons stops after a few seconds, all the neutron-rich and unstable nuclei stranded at different points along the r-process path return to the path of stability by β decays as depicted by the set of parallel and diagonally-directed arrows in the figure. Neutron rich nuclei heavier than Bi decay by both, α and β emissions, to reach stability as isotopes of the element, Pb. Whereas, for most of the stranded nuclides this adjustment process is quickly (on a geological time scale) completed, the adjustment process still goes on today for neutron-rich nuclides like ^{238}U, ^{232}Th, and ^{87}Rb. It is sobering to think that these isotopes, so fundamental to quantify the geologic time scale were produced in an extremely brief neutron burst during the explosion of massive red giant stars (Langmuir and Broecker, 2012).

3.2.8 p-process

The p-process (proton capture) also operates in supernovae and is responsible for the lightest isotopes of a given element. The probability of proton capture is much less than that of neutron capture, as the proton must have sufficient energy to overcome the coulomb or electrostatic repulsion of the target nuclide before colliding with it. So, p-process isotopes are invariably very low in abundance.

The reason that the heavy elements are rare (Figure 1.6) is because the process by which they are formed is rare; approximately only one in a million stars is massive enough to go supernovae.

4 Isotopics

> The beginning of wisdom is to call things by their right names.
>
> **Confucius**

4.1 Introduction

The basic measurement in isotope geochemistry is the quantitative variation of the relative and absolute numbers of isotopes of an element caused by natural physical and chemical processes and created artificially in the laboratory. This chapter deals with various topics related to this end.

4.2 Isotopic Abundance

The fundamental unit of chemistry is the atom or molecule. It is, therefore, necessary to be able to measure and express the number of atoms or molecules in a chemical system (natural or artificial). However, the numbers of atoms, or atomic groups (molecules), even in a very small chemical system are far too numerous to be counted. For example, just one microgram of sodium chloride (NaCl) will contain ~10^{16} sodium chloride molecules, each with one sodium and one chlorine atom. Chemists have, thus, developed a practical method to count such enormous numbers of chemical units by simply weighing them. This is based on the fundamental atomic theory that equal numbers of any atom will be contained in one gram atomic weight (atomic weight expressed in grams) of its element. This 'equal number', measured experimentally, is 6.023×10^{23}, and is called the Avogadro's number after its discoverer.

The fundamental unit of isotope chemistry is the isotope or isotopic molecule. Numbers of isotopes of an element will also be far too numerous to count, like atoms in a chemical system. So, like the gram atomic weight in chemistry, it is more convenient to use, in isotope chemistry, the unit 'mole' (gram molecular weight) defined as the amount of material which contains as many particles as there are carbon (C) atoms in exactly 12 gram of pure ^{12}C. This number is the Avogadro's number, 6.023×10^{23}. This definition emphasizes that the mole refers to a fixed number of any type of identical items (e.g. mole of electrons, mole of molecules, or even a mole of men). The product of mole and Avogadro's number gives the number of a given isotope in a chemical system, and the product of mole and atomic weight of the isotope gives the mass of that isotope in atomic mass units in the system. The definition of a mole permits the calculation of the value of the unified atomic mass unit, u.

Mass of one mole of ^{12}C atoms = N_A × mass of one ^{12}C atom = N_A × 12 u = 12 g

Hence, 1 u = 1.660×10^{-24} g. In the rest of the book the symbol for an isotope will also be used to designate its abundance in absolute numbers or moles, unless specified otherwise.

4.3 Isotope Effect in the Nuclear and Atomic Domains

Isotopes of an element differ only in their neutron number. This difference in neutron population of a nuclide results in substantial difference in its nuclear properties such as spin, parity, statistics, structure, nuclear reaction rate or probability, binding energy, and stability/instability with time (Kaplan, 1955; Leighton, 1959). In the atomic domain the difference in neutron number only results in the systematic variation of the overall mass among isotopes. If an isotope is radioactive, it will decrease in its abundance with time and there will be a complementary increase in the abundance of only one isotope (radiogenic) of the daughter element associated with it. This is the basis of radiometric dating. On the other hand, the systematic difference in the overall mass of the isotopic atoms produces subtle, but significant differences in their chemical, spectroscopic, thermodynamic, and kinetic properties. This results in nonequilibrium (kinetic) and equilibrium fractionation of all isotopes during physical, chemical, and biological processes (Criss, 1999). As the fractional mass difference between isotopes is large in light elements like hydrogen (H), C, nitrogen (N), oxygen (O) and sulphur (S), isotope fractionation is more pronounced in them relative to medium and heavy elements. This isotope fractionation forms the foundation of the major field of stable isotope geochemistry with applications in a variety of fields. It must be noted that isotope specific variations in elements, as studied in radiogenic isotope geochemistry, are directly nuclear in origin, whereas, isotope mass-dependent variations as investigated in stable isotope geochemistry are ultimately nuclear in origin (Criss, 1999; Sharp, 2007). Stable isotope geochemistry will not be considered further in this book.

4.4 Notation of Isotopic Abundances

Consider an element made up of a mixture of (say) four isotopes, a, b, c, and d with their abundances in moles also given by the same symbols and their masses given by m_a, m_b, m_c, and m_d, respectively.

Fractional isotopic abundance of a, F_a: $\dfrac{a}{(a+b+c+d)}$ (4.1)

Percentage abundance of a: $F_a \times 100$ (4.2)

Isotopic ratio of a to b, R^{ab}: $\left(\dfrac{a}{b}\right)$ (4.3)

Inter-relationship between F_a and R^{ab}

$$F_a = \frac{a}{(a+b+c)} \quad \text{for three isotopes} \tag{4.4}$$

$$= \frac{1}{\left[1+\left(\dfrac{b}{a}\right)+\left(\dfrac{c}{a}\right)\right]} = \frac{1}{(1+R^{ba}+R^{ca})} \tag{4.5}$$

Atomic weight is the sum of isotopic masses in amu, weighted according to their abundance.

$$\text{Atomic weight} = \frac{(am_a + bm_b + cm_c + dm_d)}{(a+b+c+d)}$$

$$= \frac{\left[1m_a + \left(\dfrac{b}{a}\right)m_b + \left(\dfrac{c}{a}\right)m_c + \left(\dfrac{d}{a}\right)m_d\right]}{\left[1+\left(\dfrac{b}{a}\right)+\left(\dfrac{c}{a}\right)+\left(\dfrac{d}{a}\right)\right]} \tag{4.6}$$

For example: For strontium (Sr), $\left(\dfrac{84}{86}\right) = 0.056584$, $\left(\dfrac{87}{86}\right) = 0.71025$, $\left(\dfrac{88}{86}\right) = 8.37521$.

$$\text{Atomic weight} = \frac{\begin{bmatrix}(0.056584 \times 83.90889 + 1 \times 85.909273 \\ + 0.71025 \times 86.90889 + 8.37521 \times 87.905625)\end{bmatrix}}{(0.056584 + 1 + 0.71025 + 8.37521)}$$

$$= 87.61674 \text{ u}.$$

Note: If $\left(\dfrac{87}{86}\right)$ is different from 0.71025, it must be substituted.

$$\text{Mass fraction of } ^{86}\text{Sr} = \frac{bm_b}{(am_a + bm_b + cm_c + dm_d)}$$

$$= \frac{m_b}{\left[\left(\dfrac{a}{b}\right)m_a + 1m_b + \left(\dfrac{c}{b}\right)m_c + \left(\dfrac{d}{b}\right)m_d\right]} \tag{4.7}$$

$$= \frac{85.909184}{888.61251} = 0.096678$$

$$\text{Mass fraction of } ^{87}\text{Rb} = \frac{86.909184}{(2.59265 \times 84.91180 + 1 \times 86.909184)} = 0.28304$$

$$\left(\dfrac{^{87}\text{Rb}}{^{86}\text{Sr}}\right)(\text{wt. ratio}) = \left(\dfrac{0.28304}{0.096678}\right)\left(\dfrac{\text{Rb}}{\text{Sr}}\right)(\text{wt. ratio})$$

Small changes in isotopic composition are usually reported relative to that of a selected standard measured on the same mass spectrometer with the same machine bias (Chapter 6). This technique of differential comparison has been the cornerstone of precise isotopic analysis since it was introduced around 1950 in Harold Urey's laboratory at the University of Chicago.

For example: Let $R_x^{ab} = 0.3663$ and $R_s^{ab} = 0.3659$ be the ratios of isotope a to isotope b in the standard, x, and sample, s, respectively.

$$\text{Deviation from standard in parts per 100} = \left[\left(\frac{R_s^{ab}}{R_x^{ab}}\right) - 1\right] 100 \tag{4.8}$$

$$\text{Deviation in parts per 1000, unit } \delta = \left[\left(\frac{R_s^{ab}}{R_x^{ab}}\right) - 1\right] 1{,}000 \tag{4.9}$$

$$\text{Deviation in parts per 10,000, unit } \varepsilon = \left[\left(\frac{R_s^{ab}}{R_x^{ab}}\right) - 1\right] 10{,}000 \tag{4.10}$$

Machine bias is either arranged to cancel exactly between standard and sample, or is corrected internally or externally before comparison (Chapter 6).

4.5 Mixtures of Isotopically Different Components

Mass balance calculations are of general importance in isotopic studies (Albarede, 2003). Examples include: (1) calculation of isotopic abundances in mixtures of isotopically differing materials or components, (2) isotope dilution analysis, (3) the correction of experimental results for the effects of blanks (contamination), and (4) external correction of machine bias (Chapter 6) using a double spike.

Consider a mixture of many components, say i.

Species (isotopes/elements) in each: a, b, c, d, etc. (excluding i and m).

Species concentration (in moles) in the ith component: a_i, b_i, c_i, d_i, etc.

Species content in ith component: A_i, B_i, C_i, D_i, etc.

Mass of the ith component: M_i

Mass of the mixture: M_m

Species in the mixture: a_m, b_m, c_m, d_m, etc.

Species content in the mixture: A_m, B_m, C_m, D_m

4.5.1 General multicomponent mixture

$$M_m = M_1 + M_2 + M_3 + M_4 + \cdots + M_i \tag{4.11}$$

$$M_m a_m = M_1 a_1 + M_2 a_2 + M_3 a_3 + \cdots + M_i a_i \tag{4.12}$$

$$M_m b_m = M_1 b_1 + M_2 b_2 + M_3 b_3 + \cdots + M_i b_m$$ \tag{4.13}$$

$$A_i = M_i a_i$$

Isotopics

$$a_m = \left(\frac{M_1}{M_m}\right)a_1 + \left(\frac{M_2}{M_m}\right)a_2 + \left(\frac{M_3}{M_m}\right)a_3 + \cdots + \left(\frac{M_i}{M_m}\right)a_i$$

$$= f_1 a_1 + f_2 a_2 + f_3 a_3 + \cdots + f_i a_i \tag{4.14}$$

where, $\Sigma f_i = f_1 + f_2 + f_3 + \cdots + f_i = 1$ (4.15)

Equation (4.15) is called the closure condition; f is the mixing parameter and represents the fraction of the component mass in the total mixture.

4.5.2 Ratio of two species in a multicomponent mixture

$$a_m = f_1 a_1 + f_2 a_2 + f_3 a_3 \cdots + f_i a_i$$

$$b_m = f_1 b_1 + f_2 b_2 + f_3 b_3 \cdots + f_i b_i$$

$$b_m \left(\frac{a}{b}\right)_m = f_1 b_1 \left(\frac{a}{b}\right)_1 + f_2 b_2 \left(\frac{a}{b}\right)_2 + f_3 b_3 \left(\frac{a}{b}\right)_3 \cdots + f_i b_i \left(\frac{a}{b}\right)_i$$

$$\left(\frac{a}{b}\right)_m = \left(\frac{f_1 b_1}{b_m}\right)\left(\frac{a}{b}\right)_1 + \left(\frac{f_2 b_2}{b_m}\right)\left(\frac{a}{b}\right)_2 + \left(\frac{f_3 b_3}{b_m}\right)\left(\frac{a}{b}\right)_3 + \cdots + \left(\frac{f_i b_i}{b_m}\right)\left(\frac{a}{b}\right)_i$$

$$= \phi_b^1 \left(\frac{a}{b}\right)_1 + \phi_b^2 \left(\frac{a}{b}\right)_2 + \phi_b^3 \left(\frac{a}{b}\right)_3 + \cdots \phi_b^i \left(\frac{a}{b}\right)_i \tag{4.16}$$

where, $\Sigma \phi_b^i = \left(\frac{f_1 b_1}{b_m}\right) + \left(\frac{f_2 b_2}{b_m}\right) + \left(\frac{f_3 b_3}{b_m}\right) + \cdots + \left(\frac{f_i b_i}{b_m}\right) = \left(\frac{1}{b_m}\right)(f_1 b_1 + f_2 b_2 + f_3 b_3 + \cdots) = 1$ (4.17)

$\Sigma \phi_i^b$ represents the fraction of species b residing in phase i. Equation (4.17) is a closure condition equivalent to Equation (4.15), but is restricted to species b. Weights f_i are assigned to concentration balance. Weights ϕ_i are assigned to ratio balance.

4.5.3 Mixture of two components with only one species

Closure condition, $f_1 + f_2 = 1$

Mass balance, $a_m = f_1 a_1 + f_2 a_2 = f_1 a_1 + (1 - f_1) a_2$ (4.18)

a_m is the concentration (in moles) of a conservative species in a two component mixture.

4.5.4 Mixture of two components with two species (a_1, b_1) and (a_2, b_2)

$$a_m = f_1 a_1 + (1 - f_1) a_2 \tag{4.19}$$

$$b_m = f_1 b_1 + (1 - f_1) b_2 \tag{4.20}$$

$$f_1 = \frac{(a_m - a_2)}{(a_1 - a_2)} = \frac{(b_m - b_2)}{(b_1 - b_2)}$$

Then, $$a_m = \left[\frac{(a_1 - a_2)}{(b_1 - b_2)}\right] b_m + \frac{(a_2 b_1 - a_1 b_2)}{(b_1 - b_2)} \tag{4.21}$$

Species vs species plot is linear, like $y = mx + c$.

4.5.5 Ratio of two species vs concentration of one species

Dividing Equation (4.21) by b_m gives:

$$\left(\frac{a}{b}\right)_m = \frac{(a_2 b_1 - a_1 b_2)}{(b_1 - b_2)\left(\frac{1}{b_m}\right)} + \frac{(a_1 + a_2)}{(b_1 - b_2)} \tag{4.22}$$

Plot of $\left(\dfrac{a}{b}\right)_m$ vs $\left(\dfrac{1}{b_m}\right)$ is linear.

4.5.6 Ratio (a/b) vs ratio (c/d) of species a, b, c, d in two components

Let $y_1 = \left(\dfrac{a_1}{b_1}\right)$, $y_2 = \left(\dfrac{a_2}{b_2}\right)$, and $y_m = \left(\dfrac{a_m}{b_m}\right)$ \hfill (4.23)

$x_1 = \left(\dfrac{c_1}{d_1}\right)$, $x_2 = \left(\dfrac{c_2}{d_2}\right)$, and $x_m = \left(\dfrac{c_m}{d_m}\right)$ \hfill (4.24)

From equations (4.19) and (4.20),

$$y_m = \frac{[a_1 f + a_2(1-f)]}{[b_1 f + b_2(1-f)]} \tag{4.25}$$

$$= \frac{[b_1 y_1 f + b_2 y_2 (1-f)]}{[b_1 f + b_2(1-f)]}$$

which gives f as:

$$f = \frac{(b_2 y_2 - b_2 y_m)}{[(b_1 - b_2) y_m - b_1 y_1 + b_2 y_2]} \tag{4.26}$$

$$\frac{f}{(1-f)} = \frac{(b_2 y_2 - b_2 y_m)}{(b_1 y_m - b_1 y_1)} \tag{4.27}$$

Similarly,

$$x_m = \frac{[d_1 x_1 f + d_2 x_2 (1-f)]}{[d_1 f + d_2 (1-f)]}$$

$$\frac{f}{(1-f)} = \frac{(d_2 x_2 - d_2 x_m)}{(d_1 x_m - d_1 x_1)} \tag{4.28}$$

Equating the right side of equations (4.27) and (4.28) and simplifying give

$(b_2 d_1 y_1 - b_1 d_2 y_2) x_m + (b_1 d_2 - b_2 d_1) x_m y_m + (b_2 d_1 x_1 - b_1 d_2 x_2) y_m + (b_1 d_2 x_2 y_1 - b_2 d_1 x_1 y_2) = 0$ \hfill (4.29)

This is the general equation to a hyperbola in the form: $A x_m + B x_m y_m + C y_m + D = 0$ \hfill (4.30)

where, $\quad A = b_2 d_1 y_1 - b_1 d_2 y_2$

$$B = b_1d_2 - b_2d_1$$
$$C = b_2d_1x_1 - b_1d_2x_2$$
$$D = b_1d_2x_2y_1 - b_2d_1x_1y_2$$

This equation is identical to Equation 11 in Vollmer (1976) and Equation 2 in Langmuir et al. (1977), if (a/b) is changed to (u/a), and (c/d) to (v/b).

$$(a_2b_1y_1 - a_1b_2y_2)x_m + (a_1b_2 - a_2b_1)x_my_m + (a_2b_1x_1 - a_1b_2x_2)y_m + (a_1b_2x_2y_1 - a_2b_1x_1y_2) = 0 \qquad (4.31)$$

The curvature of this mixing curve is determined by the B coefficient. If a parameter r is defined as $(b_1/b_2)/(d_1/d_2)$, its value determines the curvature of the hyperbola. If $r = 1$ ($b_1d_2 = b_2d_1$), the second term is zero, and the equation becomes a straight line in a plot of y_m vs x_m. As r progressively differs above and below unity, the mixing curve becomes more pronounced, as shown (not to scale) in Figure 4.1.

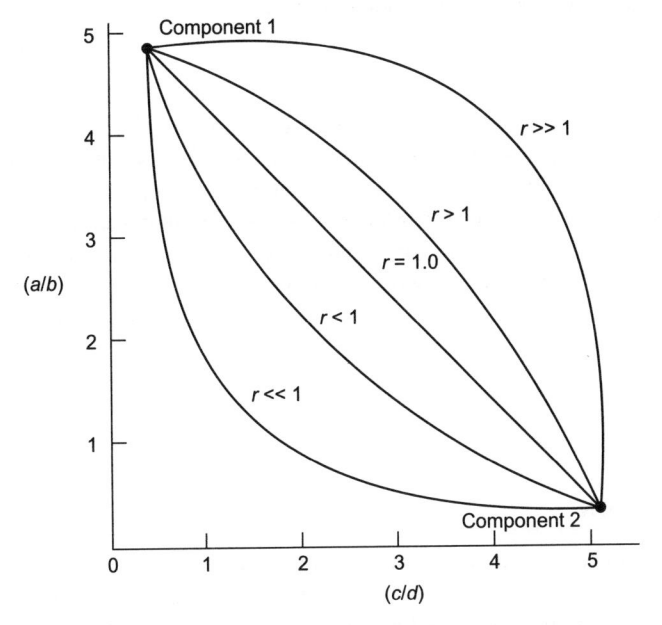

Figure 4.1 Theoretical mixing curves between two components. The curvature of the hyperbola is controlled by a parameter defined in the text

For ratio vs concentration plot, $d_1 = d_2 = 1$ and $B = b_1 - b_2$. It will also be a hyperbola unless $b_1 = b_2$.

For element vs element plot, $b_1 = b_2 = d_1 = d_2 = 1$, $B = 0$, and the equation is a straight line.

5 Radioactivity and Radiometric Dating

Liquids and gases forget, but rocks remember.

J A O'Keefe

...use of isotope data for geological purposes – measuring of ages, determining the sources of basalts, or understanding the evolution of the mantle – is no different from similar use of other geological, petrological, geochemical, or geophysical data. In all cases, the link between experimental data and geological interpretation must be continually forged. There will be no cookbooks or black boxes which will permit us to grind out generally valid geological conclusions.

Wetherill et al. (1981)

5.1 Introduction

Radioactivity is the spontaneous transformation or decay of a potentially unstable nuclide (a specific proton-neutron combination) into a more stable nuclide (Chapter 2). A decaying (radioactive) nuclide and its product (radiogenic) nuclide are conventionally labelled parent (p) and daughter (d), respectively. If d is radioactive, it decays to another nuclide until a stable daughter isotope is produced. As the decay of a radionuclide is a statistical phenomenon (Chapter 2), it is not possible to predict when exactly it will occur. However, it is experimentally found that in a large collection, p of radioactive nuclides at time τ, a constant fraction of it will decay in unit time (Rutherford, 1906). Using the symbols p and d to also represent the concentrations of parent and daughter isotopes, respectively, the rate of decay of the parent is expressed mathematically by the equation,

$$\frac{dp}{d\tau} = -\lambda p \qquad (5.1)$$

where, λ is the constant of proportionality, and considered the probability of decay of a single nuclide in unit time regardless of its previous history and present circumstances. Called the decay constant,

---

[*]A significant part of this chapter has already been published by the author (Gopalan, 2015).

Radioactivity and Radiometric Dating

λ is unique to each radioactive species, and does not show detectable changes due to variations in pressure, temperature, and chemical environments in planetary contexts. However, very minor changes in decay constant with pressure have been reported in rare decay systems (Dickin, 2005). It is measured in units of reciprocal time. Decay constants range from zero for a stable nuclide to 1.0 for an isotope that decays instantly. The negative sign in the equation reflects the decrease in p with time measured positively into the future. This equation can be written in terms of the fractional decrease (dp/p) in a small time interval, $d\tau$ as:

$$\left(\frac{dp}{p}\right) = -\lambda d\tau \tag{5.2}$$

If a concentration of p_1 at time τ_1 decreases to p_2 a time τ_2 later, Equation 5.2 can be integrated between these limits to give:

$$p_2 = p_1 \exp -\lambda(\tau_2 - \tau_1) \tag{5.3}$$

The complementary growth of the daughter isotope, d can be similarly derived from its production rate:

$$\frac{d(d)}{d\tau} = \lambda p$$

But it is simpler to get it as the difference between p_1 and p_2 as:

$$d = p_1 - p_2 = p_1[1 - \exp -\lambda(\tau_2 - \tau_1)] = p_2[\exp \lambda(\tau_2 - \tau_1) - 1] \tag{5.4}$$

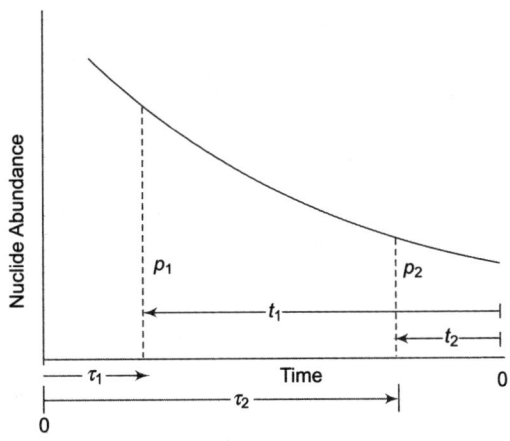

Figure 5.1 Spontaneous variation in the abundance of a radioactive isotope with time measured positively into the future and into the past. (Gopalan, 2015)

The graphical representation of Equation (5.3) is shown in Figure 5.1. Note, p_1 and p_2 are only instantaneous values in a smoothly and spontaneously decreasing concentration. The fractional decrease, (p_2/p_1) depends both on λ and the time interval ($\tau_2 - \tau_1$). The fractional decrease of a single species will depend only on the time elapsed. If a radionuclide decays in two modes with decay constants, λ_1 and λ_2 into two different daughter isotopes d^1 and d^2, respectively, Equation (5.4) gives d^1 as:

$$d^1 = \left[\frac{\lambda_1}{(\lambda_1 + \lambda_2)}\right] p_2 \{\exp[(\lambda_1 + \lambda_2)(\tau_2 - \tau_1)] - 1\} \tag{5.4a}$$

Equation (5.3) relates the surviving number in a collection of radioactive nuclides at two different times. Multiplying each side by λ gives:

$$\lambda p_2 = \lambda p_1 \exp - \lambda(\tau_2 - \tau_1) \tag{5.5}$$

This equation relates the disintegration rates [Equation (5.1)] of the surviving nuclides at two different times. The products, λp_1 and λp_2 are called activities, and are usually denoted as $[p_1]$ and $[p_2]$, respectively. Units of activity are: the Curie corresponding to 3×10^{10} disintegrations per second, and the Becquerel to one disintegration per second. If λ is large [$\gg 1/(\tau_2 - \tau_1)$], $[p_1]$ will decrease to $[p_2]$ on a laboratory time scale (minutes, hours, or days). λ can then be determined from Equation (5.5) by measuring the decrease in activity using a nuclear radiation counter over a precisely clocked time interval. If λ is small [$\ll 1/(\tau_2 - \tau_1)$], the activity will vary little over laboratory time scales. In that case, the number of disintegrations per unit time from a precisely known quantity of a radionuclide can be used in Equation (5.1) to determine λ. Very small decay constants are not precisely known because of the experimental difficulties in such absolute measurements.

In Equation (5.3), time is the independent variable and fractional decrease the dependent variable. This is inverted in the following equation:

$$(\tau_2 - \tau_1) = \left(\frac{1}{\lambda}\right) \ln\left(\frac{p_1}{p_2}\right) \tag{5.6}$$

where ln is the natural logarithm. This equation shows that:

1. Fractional decrease is equal in equal time intervals. The time interval for 50 per cent decrease in concentration is called the half-life (T) of the radionuclide.

$$T = \left(\frac{1}{\lambda}\right) \ln 2 = \frac{(0.693)}{\lambda}$$

   Half-life is an easily comprehended measure of the longevity of a nuclear species. Another, but less frequently used measure of longevity, is the mean life, Γ, defined as the average life expectancy of a radionuclide. It is equal to $(1/\lambda)$.

2. Radioactivity can be used as a chronometer or clock for measurement of time (time interval), uniquely and absolutely. The useful range of this nuclear chronometer is determined by its half-life, and is usually not more than 10 half-lives. A radioisotope with a sufficiently long half-life will, therefore, be required to measure geologic time (thousands to millions of years).

3. This chronometer can measure time intervals in the future, as well as in the past. The latter capability is of great significance in essentially integrative and historical sciences like geology and archaeology.

As it is more convenient in geology to measure time positively into the past, Equation (5.6) can be rewritten for this purpose as:

$$(t_1 - t_2) = \left(\frac{1}{\lambda}\right) \ln\left(\frac{p_1}{p_2}\right) \tag{5.7}$$

where, $t_1 > t_2$ (Figure 5.1). From now on time measured forward will be denoted by 'τ', and time measured positively into the past by 't'. Unless otherwise stated, the end of the decay interval will, hereafter, be the present time, that is, $t_2 = 0$. Writing t_1 as t, p_1 as p_t to represent the concentration of radionuclide at time, t in the past, and p_2 as p_o to represent the present day value, the decay equation becomes:

$$p_o = p_t \exp(-\lambda t)$$

The right suffix 'o' to indicate quantities at the present time will, hereafter, be omitted, as in Equation (5.8), primarily to avoid possible confusion due to this suffix being used in literature to refer to initial values.

$$p = p_t \exp(-\lambda t) \quad (5.8)$$

Time intervals, in general, are usually given in units of kilo-years (Ky), mega-years (My), and giga-years (Gy). Time intervals ending at the present (with the conventional meaning of age) are given in units of kilo-annum (Ka), mega-annum (Ma), and Giga-annum (Ga).

The exponential function, $\exp(-\lambda t)$ can be expanded into an infinite power series as:

$$\exp(-\lambda t) = \frac{1+(-\lambda t)}{1!} + \frac{(-\lambda t)^2}{2!} + \frac{(-\lambda t)^3}{3!} + \cdots + \frac{(-\lambda t)^n}{n!} + \cdots$$

where, the factorial, $n! = n(n-1)(n-2) \ldots 3.2.1$. For $\lambda t \ll 1$, Equation (5.8) becomes linear as:

$$p = p_t(1 - \lambda t)$$

5.2 RADIOACTIVE AND RADIOGENIC ISOTOPE DATING

If a radioactive decay interval is to be related to the chronology of a natural system or object, it is necessary that the former is linked directly or explicitly to the latter. For example, if a collection of radioisotopes with an appropriate decay constant is incorporated in a natural system at the time of its formation, the ratio of the original number to the present number in the collection will represent the 'age' of the natural system. Radioactive nuclides can be introduced into a natural object by natural, physical, or chemical processes in two main ways:

1. Incorporation of a chemical element with its mixture of radioactive and stable isotopes. As isotopes are largely insensitive to chemical fractionation, the isotopic composition of the incorporated elements will be the same as in their parent source. Except in some special cases, some daughter element will also be incorporated. The extent of fractionation or discrimination between the parent and daughter elements will depend on their chemical contrast.

2. In situ production of only the radioactive parent isotope by natural nuclear reactions (Lal and Peters, 1967; Lal, 1988) or from the decay of a preexisting radioactive parent (Ku, 1976).

So, the isotopic composition of a piece of natural sample can, in principle, be related to the chronology of the major processes and events that led to its formation or re-formation. But the problem is that we can measure the isotopic composition or other attributes, like chemical makeup, mineralogical composition, and residual magnetism of a sample, only as it is available to us at present. We can have no idea which isotope composition to look for in the first place, and whether such composition was smeared or partially erased or overprinted by changes and alterations suffered by a sample, since its original formation. Under the circumstances, we have to first characterize the sample under study chemically, mineralogically, and texturally to get some idea of what isotopic memory it is likely to retain, and also its significance in the larger context of geology. We can then construct plausible physical and chemical models to extrapolate the observed data into the past with the purpose of describing the time and mechanism of formation, and the parental source of the object. As stated succinctly by Wasserburg (1987), all natural objects represent a library of experiments that have already been done at different times, places, and conditions. A unique interpretation of the data may not be possible in each case. However, our effort must be to extract a maximum quantity of reasonably certain knowledge by an intelligent selection of samples. An admixture of healthy skepticism and imaginative creativity in the selection of samples is the key to important scientific insights.

Let us consider a parcel of natural matter of arbitrary size that is available to us at present, and represented by the rectangular box on the right in Figure 5.2. It will contain many elements and also other characteristic features. But, we focus only on one each of the radioactive isotopes (p), and non-radioactive (stable) isotopes (p') of the parent element, and one each of the radiogenic isotopes (d), and non-radiogenic isotopes (d') of its daughter element. We will assume that the corresponding isotopes were p_t, p_t', d_t, and d_t' in the same sample at some time, t, in the past in the box on the left. 't' may be any time in the past, but not exceeding the time since the parent-daughter ratio in the box changed only due to radioactive decay. The quantities in the left box can be uniquely related to the present quantities in the right box, if the parcel of matter within the box has remained chemically closed (at least with respect to the two elements under consideration) over the time interval of interest, t, as:

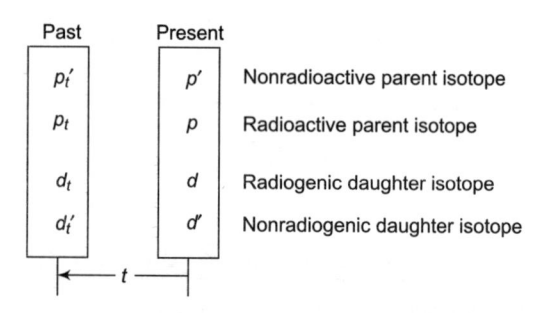

Figure 5.2 Abundances of parent and daughter isotopes at present in a chemical system (box on the right) and their projections in the same system at an earlier time t under chemically closed condition (box on the left). (Gopalan, 2015)

$p_t' = p'$ Time-invariant stable isotope of the parent

$p_t = p \exp(\lambda t)$ Radioactive decay with time

$d_t = d - (p_t - p)$ Initial radiogenic isotope of the daughter element

$d_t' = d'$ Time-invariant non-radiogenic isotope of the daughter

The following different cases are conceivable and indeed met with in natural systems.

5.2.1 p is measurable and measured, p_t is known or assumed, and p', d and d' not measured

The basic radioactive decay equation, $p = p_t \exp(-\lambda t)$ gives t directly.

Radioactivity and Radiometric Dating

5.2.2 Both p and p' are measurable and measured, and p_t is known or assumed

Division of the radioactive decay equation by a time invariant nonradioactive parent isotope ($p' = p_t'$) gives:

$$\left(\frac{p}{p'}\right) = \left(\frac{p}{p'}\right)_t \exp(-\lambda t),$$

eliminating the need for measuring absolute concentrations.

5.2.3 p is measurable and measured, p_t is not known, and d' is absent

As absence of d' means the absence of d_t also:

$$d = (p_t - p) = p[\exp(\lambda t) - 1]. \tag{5.9}$$

t can be calculated from only the residual parent and accumulated daughter.

5.2.4 p, d, and d' are measurable and measured, whereas, p_t and d_t are unknown

This most general case is governed by the relation:

$$(d - d_t) = (p_t - p) = p[\exp(\lambda t) - 1]. \tag{5.10}$$

Dividing this equation throughout by any constant quantity is mathematically trivial. However, division by one particular constant quantity, namely $d_t' = d'$ of the non-radiogenic daughter isotope leads to a conceptually profound link between geochemistry and radioactivity, besides eliminating the need for absolute measurements.

$$\left(\frac{d}{d'}\right) = \left(\frac{d}{d'}\right)_t + \left(\frac{p}{d'}\right)[\exp(\lambda t) - 1]. \tag{5.11}$$

The chemical process that created this object could have altered the relative proportions of the parent and daughter elements because of their chemical contrasts, but not their isotopic compositions relative to its immediate precursor or source. So, the first term on the right side of this equation is a tracer of its previous history, and the second term the time that has elapsed since its formation with a non-zero parent-daughter ratio. Of course, many other observations on the sample, like chemical composition, mineralogy, and texture, are also relevant to its past history, but not so explicitly and quantitatively as the isotopic composition. This is the fundamental age equation. It has been used widely and creatively in one way or another, whenever chronological conclusions are based on decay of a radioactive parent to a stable daughter isotope.

5.2.5 p is zero, but known to be short-lived (1/λ << t) from its daughter d. p' is measurable and so are d and d'

Since all parent isotopes would have decayed into their daughter isotopes:

$$d - d_t = p_t$$

As $d' = d_t'$ and $p' = p_t'$, this equation can be rewritten as:

$$\left(\frac{d}{d'}\right) = \left(\frac{d}{d'}\right)_t + \left(\frac{p}{p'}\right)_t \left(\frac{p'}{d'}\right) \tag{5.12}$$

A direct connection of presently measured (d/d') with time, as measured backward from the present, has disappeared. But, if (d/d') is measurably higher than (d/d')$_t$ from additional information or linear with presently measurable (p'/d'), the former presence of a short-lived isotope in the object is indicated. This case refers to the use of 'fossil' or 'extinct' radioisotopes for insights into the earliest history of the Solar System and the earth (chapters 9 and 11).

5.2.6 p is long-lived, but both p and p' are missing in the sample

If (d/d') is significantly higher than possible (d/d')$_t$ values, the implication is that the association of the parent element with the sample lasted less than t. The growth of the daughter until the withdrawal (for whatever reason) of its parent at time t_1 in the past will be given by:

$$d - d_t = (p_t - p_{t1}) = [p_t - p_t.\exp{-\lambda(t - t_1)}],$$

which simplifies, after division by d', to:

$$\left(\frac{d}{d'}\right) = \left(\frac{d}{d'}\right)_t + \left(\frac{p_t}{d'}\right)[1 - \exp{-\lambda(t - t_1)}] \tag{5.13}$$

This equation can be solved for t_1 only if the initial parent-daughter ratio and t are known. This case pertains to radiogenic isotope composition of elements not associated at present with any of their respective parent elements in natural samples, for example, lead (Pb) with negligible uranium (U) as in galena, and strontium (Sr) with negligible rubidium (Rb) as in carbonates.

5.2.7 Radiogenic daughter isotope, d is not stable, but radioactive

The growth of the daughter, in this case, can be derived from first principles following Wetherill et al. (1981). The concentration of the radioactive daughter, d, produced during a small interval of time $\Delta t'$ at time t' in the past ($t < t' < 0$) with a present day concentration p of its parent will, according to the basic laws of radioactive decay, be:

$$\Delta d_{t'} = (\Delta t' \lambda_p p) \exp(\lambda_p t')$$

where, the present concentration, p of the parent is augmented by the factor of $\exp(\lambda_p t')$ to represent its concentration at time t' in the past, and λ_p is its decay constant. During the interval since t', this batch of $\Delta d_{t'}$ will decay to:

$$\Delta d = (\Delta t' \lambda_p p) \exp(\lambda_p t') \exp(-\lambda_d t') = (\Delta t' \lambda_p p) \exp(\lambda_p - \lambda_d)t' \tag{5.14}$$

where, λ_d is the decay constant of the daughter nuclide. The total concentration from all batches produced during the time interval t will be given by integrating the above equation between the time limits 0 and t:

$$d = \lambda \int p \exp(\lambda_p - \lambda_d) t' dt'$$

$$= \left[\frac{\lambda_p}{(\lambda_p - \lambda_d)}\right] (p)[\exp(\lambda_p - \lambda_d)t - 1] \tag{5.15}$$

This represents the quantity of the daughter that has formed by the decay of its parent, but has not yet decayed. If the initial concentration of the daughter is d_I, it will decay to $d_I.\exp(-\lambda_d t)$ during the interval t. So, the net daughter concentration now will be:

$$d = d_t \exp(-\lambda_d t) + \left[\frac{\lambda_p}{(\lambda_p - \lambda_d)}\right] (p)[\exp(\lambda_p - \lambda_d)t - 1] \qquad (5.16)$$

If the daughter is stable as in previous cases, $\lambda_d = 0$, the equation reduces to:

$$d = d_t + p\,[\exp(\lambda_p t) - 1],$$

as is to be expected. However, in the opposite case when $\lambda_p \ll \lambda_d$, the equation changes to:

$$d = d_t \exp(-\lambda_d t) + \left(\frac{\lambda_p p}{\lambda_d}\right)[1 - \exp(-\lambda_d t)] \qquad (5.17)$$

As t increases, d will increase initially, but approach a constant value of $(\lambda_p p/\lambda_d)$ for $t \gg 1/\lambda_d$. Thus, the increase of d due to the decay of p will be useful for measuring time intervals $t \sim 1/\lambda_d$ only. When $t \gg 1/\lambda_d$, $\lambda_d d = \lambda_p p$, that is, the rate of decay of the radioactive daughter (its activity) becomes equal to that of its long-lived parent. This condition is known as secular equilibrium.

As before, we can convert this equation between absolute quantities to one between ratios by throughout dividing by the constant quantity of any non-radiogenic daughter $d' = d'_t$ to give:

$$\left(\frac{d}{d'}\right) = \left(\frac{d}{d'}\right)_t \exp(-\lambda_d t) + \left(\frac{\lambda_p}{\lambda_d}\right)\left(\frac{p}{d'}\right)[1 - \exp(-\lambda_d t)] \qquad (5.18)$$

This provides two ways in which a short-lived daughter isotope can be used to calculate time intervals. If, in the sample, p is missing, the second term is zero.

$$\left(\frac{d}{d'}\right) = \left(\frac{d}{d'}\right)_t \exp(-\lambda_d t)$$

This corresponds to the unsupported decay of any daughter isotope separated from its parent. If p is present, but $\lambda_p p \gg \lambda_d d$ and $d_t = 0$, then:

$$\left(\frac{d}{d'}\right) = \left[\frac{\lambda_p}{(\lambda_p - \lambda_d)}\right]\left(\frac{p}{d'}\right)[1 - \exp(-\lambda_d t)]$$

This case refers to the build-up from zero of a daughter isotope from its parent isotope.

The more general scenario of a parent nuclide decaying through a chain or series of radioactive daughters before forming a stable daughter has been considered by Bateman (1910). Faure and Mensing (2005) have discussed such chain decays and shown that, eventually, the activities of all nuclides in the chain become equal. This means that the rate of growth of the stable daughter nuclide, n, is the same as the rate of decay of its ultimate parent, p. So, the age Equation (5.11) can be modified to:

$$\left(\frac{n}{n'}\right) = \left(\frac{n}{n'}\right)_t + \left(\frac{p}{n'}\right)[\exp(\lambda_p t) - 1] \qquad (5.19)$$

where, n' is a non-radiogenic isotope of the daughter element. This relation is the key to the measurement of ages of natural objects (rocks and minerals) based on the decay of a long-lived parent through a number of intermediate radioactive daughters to a final stable daughter (see next section).

In many of the above cases, calculation of t requires a measurement of the absolute or relative concentrations of both parent and daughter elements (p and d'), in addition to the isotopic composition of the latter (d/d'). A separate and explicit measurement of the parent isotope can be dispensed with, and age calculation made from only isotope compositions of the daughter element if a definite fraction of the parent element can be converted in the laboratory to a third isotope $d*$ of the daughter element. In that case, the fundamental age equation changes to:

$$\left(\frac{d}{d'}\right) = \left(\frac{d}{d'}\right)_t + \left(\frac{Cd*}{d'}\right)[\exp(\lambda t) - 1] \qquad (5.20)$$

where, C is the conversion factor of the third isotope, $d*$ into p. t can now be calculated from the measurement of only (d/d') and ($d*/d'$) ratios in the daughter element. Merrihue and Turner (1966) were the first to convert a fraction of one of the stable isotopes of potassium (K), ^{39}K, into ^{39}Ar of argon by neutron irradiation in a nuclear reactor to measure ^{40}K indirectly as ^{39}Ar. This resulted in an elegant refinement of the classical ^{40}K-^{40}Ar method into the modern ^{39}Ar-^{40}Ar method (Chapter 9). Other proposals for such proxy measurement of parent nuclide have not found wide applications (Gopalan and Rao, 1976; Mitchell, 1983).

5.3 Long-lived Parent-daughter Couples used in Radiometric Dating

Table 5.1 gives p', p, d, d', decay mode, decay constant (Steiger and Jager, 1977), and half-life for a majority of long-lived parent-daughter systems used in radiometric dating. It is instructive to consider the data in this table in the light of what we have already learnt in the previous chapters.

Table 5.1 Parent-daughter Systems with Half-lives Comparable or Longer than 10^9 Years

p' (%)	p (%)	d	d'	Mode	$\lambda (y^{-1})$	Half-life ($10^9 y$)
^{39}K(93.26)	^{40}K (0.012)	^{40}Ar	^{36}A	ec	0.58×10^{-10}	11.9
^{39}K	^{40}K	^{40}Ca	^{44}Ca	β^-	4.96×10^{-10}	1.39
^{85}Rb(72.16)	^{87}Rb(27.84)	^{87}Sr	^{86}Sr	β^-	1.42×10^{-11}	48.8
^{144}Sm(3.1)	^{147}Sm(15.07)	^{143}Nd	^{144}Nd	α	6.54×10^{-12}	106
^{175}Lu(97.4)	^{176}Lu(2.6)	^{176}Hf	^{177}Hf	β^-	2.0×10^{-11}	35
^{176}Lu	^{176}Yb	ec	–	–		
^{185}Re(36.07)	^{187}Re(63.93)	^{187}Os	^{188}Os	β^-	1.5×10^{-11}	46
–	^{232}Th (100)	^{208}Pb	^{204}Pb	$6\alpha, 4\beta^-$	4.95×10^{-11}	14
–	^{235}U (0.73)	^{207}Pb	^{204}Pb	$7\alpha, 5\beta^-$	9.85×10^{-10}	0.70
–	^{238}U (99.27)	^{206}Pb	^{204}Pb	$8\alpha, 6\beta^-$	1.55×10^{-10}	4.46

The half-lives of the parent isotopes vary widely. Relative to the age of the Solar System or the earth (~4.5×10^9 years, as we will see later), the half-life of ^{235}U is more than six times shorter, that of ^{238}U is about the same, and that of ^{147}Sm of samarium more than 25 times longer. This means that the original inventory of ^{235}U in the earth is depleted now by about 99 per cent, that of ^{238}U by ~50 per cent, and that of ^{147}Sm by only a few per cent. The present-day ratio of the two isotopes of U, ^{235}U/^{238}U is, therefore, quite low at 1/137.8 (or 0.007), which translates into a ^{235}U abundance of 0.72 per cent. It is easy to calculate this ratio at any time, t, in the past using the basic radioactive decay equation, as:

$$\left(\frac{1}{137.8}\right) = \left(\frac{^{235}U}{^{238}U}\right)_t \left[\frac{\exp(-\lambda^{235}t)}{\exp(-\lambda^{238}t)}\right] \qquad (5.21)$$

Using the appropriate decay constants in the table, the ratio comes out as ~0.04 at t = 2000 Ma, and ~0.31 at t = 4500 Ma, respectively. The obvious implication of this equation is that the abundance of ^{235}U was comparable to that of ^{238}U very early in the history of the earth. A less obvious, but very significant and testable implication is that the present-day ^{235}U/^{238}U ratio should be uniform everywhere, because this ratio is assumed to have been uniform throughout the Solar System in the very beginning. So, it came as a surprise in 1972 that a routine isotopic analysis of a batch of U ores from the Oklo deposit in Gabon (a former French colony in the west equatorial Africa) revealed a ^{235}U content of 0.717 per cent, slightly, but distinctly lower than the expected value of 0.720 per cent. This deficiency was later found to be substantial—up to 0.44 per cent—in other ore bodies mined from the Oklo deposit. A possible clue to this puzzle was soon found in the prediction as early as 1953 by G W Wetherill (then a 27 year old graduate student in the University of Chicago) that some U deposits might have once operated as natural versions of the artificial nuclear fission reactor realized by Enrico Fermi a decade earlier in the same university. In man-made or artificial nuclear reactors using U as a fuel, a ^{235}U nucleus is induced (Chapter 2) by a stray thermal (low energy) neutron to fission (split) into two parts, releasing energy (~200 MeV) and more than one (2.5 on the average) neutron. These secondary neutrons are slowed down (moderated) and conserved sufficiently to induce fission in other ^{235}U nuclei in the fuel charge in a nuclear chain reaction for continuous energy generation. The U fuel will, therefore, be gradually depleted in its ^{235}U content with time, whereas, its ^{238}U content will remain practically unchanged. Following Wetherill's suggestion, Paul K Kuroda, at the university of Arkansas, calculated the conditions for a natural U deposit to spontaneously undergo self-sustained fission reactions, like the minimum size of the deposit, its U concentration, abundance of ^{235}U, if natural water (the most likely ambient medium) was to moderate the fission-produced fast neutrons (Kuroda, 1956). That the Oklo deposit indeed satisfied these conditions to sustain induced fission reaction was convincingly borne out by the detection of fission products (both elements and isotopes) uniquely distinctive of such reactions. It is now known that light or natural water moderation can sustain a chain reaction in U if its ^{235}U is enriched from its natural abundance of 0.720 per cent to at least 3 per cent. That any natural deposit would have had such high ^{235}U abundance only before 1.7 billion years sets a strict younger time limit to chain reaction. An older limit would correspond to the time of significant oxygenation of the atmosphere to mobilize trace levels of U dispersed in igneous rocks into concentrated ore bodies. Cowan (1976) and Meshik (2005) have given a highly readable description of this fascinating Oklo nuclear reactor.

According to the mass energy relation (Section 1.4), only a very minute fraction of the ^{235}U mass that undergoes induced fission in nuclear reactors is converted to useful energy. The rest is transformed into radioactive fission fragments, each with roughly half the mass of ^{235}U atom, and is generally known as nuclear waste. Fission-produced nuclides with short half-lives decay quickly to become harmless, but those with longer half-lives persist for many thousands of years, posing a serious health hazard. The isolation of accumulated nuclear waste (many thousands of tons globally) from the biosphere until it decays to safe levels is still a tricky problem.

Before closing this brief digression into natural and engineered chain nuclear reaction, it is interesting to speculate what would happen if the half-life of ^{235}U were even slightly shorter, say ~500 My instead of ~700 My. In that case the present-day abundance of ^{235}U in natural U would be so scarce as to render it prohibitively expensive, if not impossible, to enrich U to the level required for either controlled chain reaction in a nuclear reactor or explosive nuclear reaction in an atomic bomb.

From Chapter 3, we can infer that ^{40}K was synthesized in stellar fusion reactions, ^{87}Rb, ^{147}Sm ^{176}Lu of lutetium and ^{187}Re of rhenium in either s- or r-type of neutron caption reactions, and ^{232}Th of thorium, ^{235}U, and ^{238}U in only r-process neutron capture reactions. The first five radioactive isotopes are lighter than the heaviest stable isotope, ^{209}Bi (of bismuth), and they all have at least one non-radioactive counterpart p' (^{147}Sm has six), but the heavier radioactive Th and U isotopes have none. It was noted in this chapter that these three nuclides are the only nuclides produced in the r-process path (Figures 3.4 and 3.5) that are yet to reach the stability zone. But for their long half-lives, they would have also decayed completely long ago, depriving geochronology and cosmochronology of a versatile pair of parent-daughter systems.

Next we consider the interesting diversity in the decay modes of the parent isotopes. ^{40}K exhibits branched decay by β^- emission to ^{40}Ca of calcium with a decay constant, λ_β, of 4.96×10^{-10} y^{-1} and by electron capture to ^{40}Ar with decay constant λ_{ec} of 0.581×10^{-10} y^{-1}. An extremely small fraction of ^{40}K also decays by β^+ emission to ^{40}Ar. It is impossible to predict how a given ^{40}K atom will decay, just as it is impossible to predict when it will decay. However, we can predict quite accurately what proportion of a large number of ^{40}K atoms will decay to each. $\lambda_{ec}/\lambda_\beta = 0.117$ is called the branching ratio, and indicates that about 10 per cent of ^{40}K atoms will eventually decay to ^{40}Ar. The radioactivity of ^{40}K, therefore, provides in principle, two dating schemes, namely K-Ar and K-Ca. ^{176}Lu also has a branching decay—to ^{176}Hf by β^- emission and to ^{176}Yb (of *Ytterbium*) by electron capture. However, the electron capture process is so scarce and so slow that ^{176}Lu-^{176}Yb is practically useless for dating. ^{87}Rb, although chemically similar to ^{40}K, decays only by β emission to stable ^{87}Sr. ^{187}Re is also only a β emitter, whereas, ^{147}Sm is an α emitter.

^{232}Th, ^{235}U, and ^{238}U are so far away from the stable ^{209}Bi that they cannot decay directly to a stable daughter near ^{209}Bi, but only through a succession of intermediate radioactive daughters, with some of them being necessarily α-active. For example, the transformation of one atom of ^{238}U to one atom of ^{206}Pb requires a net change in mass number of 32 and a net change in atomic number of 10. Table 5.1 shows that a combination of eight α decays and six β decays in series accomplishes this. The combinations for ^{235}U and ^{232}Th chain decays are also given in the table. The names, decay modes, and half-lives of the intermediate daughters in the two U chains are shown in Table 5.2. Note the tremendous diversity of half-lives (a few minutes to as long as 2,50,000 years) among the intermediate daughter nuclides. This means that all the intermediate nuclides in both chains will grow in a lump of pure U until they reach, in about a million years, a steady state concentration (secular equilibrium, discussed in Section 5.2.7), balancing their growth and decay rates, at which point ^{238}U may be deemed to decay directly to ^{206}Pb [Equation 5.19)]. It is doubly fortunate for geochronology

Radioactivity and Radiometric Dating

that two isotopes of U not only survived to the present, but also decay to two isotopes of one and the same daughter element. The coupling of these two chemically identical decay systems for refined geochronological applications is described later in this chapter. A radioactive isotope of an element can never decay directly into another isotope of that element. However, Table 5.2 shows that the ^{238}U chain contains two U isotopes, ^{238}U and ^{234}U, three polonium (Po) isotopes, ^{218}Po, ^{214}Po, and ^{210}Po, and three Pb isotopes, ^{214}Pb, ^{210}Pb, and ^{206}Pb. This is the consequence of an α decay followed by a β decay.

If the steady state condition in a chain (secular equilibrium) is disturbed and the intermediate daughters are separated or fractionated by a natural process because of their chemical contrasts, conditions envisaged in Section 5.2.7 apply. Any isolated daughter will begin to decay from its equilibrium concentration just prior to the disturbance, and any isolated parent will begin to build its daughter once again to equilibrium, as governed by the equations derived in that section. The U series disequilibrium dating has now become an active area of research in its own right (Ku, 1976). One application of a U series isotope to dating recent volcanism on the earth will be covered in Chapter 11.

Using the data in Table 5.2 it is straightforward to write down the working decay equations for each of the decay pairs, as follows:

K-Ar method,
$$\left(\frac{^{40}Ar}{^{36}Ar}\right) = \left(\frac{^{40}Ar}{^{36}Ar}\right)_t + \left[\frac{\lambda_{ec}}{\lambda_\beta + \lambda_{ec}}\right]\left(\frac{^{40}K}{^{36}Ar}\right)[\exp(\lambda t) - 1],$$

K-Ca method,
$$\left(\frac{^{40}Ca}{^{44}Ca}\right) = \left(\frac{^{40}Ca}{^{44}Ca}\right)_t + \left[\frac{\lambda_\beta}{\lambda_\beta + \lambda_{ec}}\right]\left(\frac{^{40}K}{^{44}Ca}\right)[\exp(\lambda t) - 1],$$

where, $\lambda = (\lambda_\beta + \lambda_{ec})$ is the total decay constant. The first factor in the second term on the right side of these two equations accounts for the branching decay of ^{40}K.

Rb-Sr method,
$$\left(\frac{^{87}Sr}{^{86}Sr}\right) = \left(\frac{^{87}Sr}{^{86}Sr}\right)_t + \left(\frac{^{87}Rb}{^{86}Sr}\right)[\exp(\lambda^{87} t) - 1]$$

Sm-Nd method,
$$\left(\frac{^{143}Nd}{^{144}Nd}\right) = \left(\frac{^{143}Nd}{^{144}Nd}\right)_t + \left(\frac{^{147}Sm}{^{144}Nd}\right)[\exp(\lambda^{147} t) - 1]$$

Lu-Hf method,
$$\left(\frac{^{176}Hf}{^{177}Hf}\right) = \left(\frac{^{176}Hf}{^{177}Hf}\right)_t + \left(\frac{^{176}Lu}{^{177}Hf}\right)[\exp(\lambda^{176} t) - 1)]$$

Re-Os method,
$$\left(\frac{^{187}Os}{^{188}Os}\right) = \left(\frac{^{187}Os}{^{188}Os}\right)_t + \left(\frac{^{187}Re}{^{188}}\right)[\exp(\lambda^{187} t) - 1]$$

Th-Pb method,
$$\left(\frac{^{208}Pb}{^{204}Pb}\right) = \left(\frac{^{208}Pb}{^{204}Pb}\right)_t + \left(\frac{^{232}Th}{^{204}Pb}\right)[\exp(\lambda^{232} t) - 1]$$

U-Pb method,
$$\left(\frac{^{207}Pb}{^{204}Pb}\right) = \left(\frac{^{207}Pb}{^{204}Pb}\right)_t + \left(\frac{^{235}U}{^{204}Pb}\right)[\exp(\lambda^{235} t) - 1]$$

Table 5.2 ^{238}U and ^{235}U Decay Series' Products and Their Half-lives

	Uranium-238 Series, includes ^{234}U Series					Uranium-235 Series			
Np									
U	^{238}U 4.5E9	^{234}U 2.5E5y				^{235}U 7.1E8y			
Pa	↓	^{234}Th 24 d ↗	↓			↓	^{231}Pa 3.3E4y		
Th	^{234}Th 24 d ↗	^{230}Th 8E4y				^{231}Th 25.5 h	↓	^{227}Th 18.7y	
Ac		↓					^{227}Ac 21.8y ↗	↓	
Ra		^{226}Ra 1600 y					↓	^{227}Ra 11.4d	
Fr		↓					^{223}Fr 21.8 m ↗	↓	
Rn		^{222}Rn 3.82 d						^{219}Rn 4.0s	
At		↓	^{218}At 2 s				↓		^{215}At 1E-4s
Po		^{218}Po 3.05 m ↗	↓	^{214}Po 1.8E-4s	↓	^{210}Po 138 d	^{215}Po 1.8E-5 ↗	↓	^{211}Po 0.5s
Bi		↓	^{214}Bi 19.7 m ↗	↓	^{210}Bi 5.0 d ↗	↓		^{211}Bi 2.15 m ↗	↓
Pb		^{214}Pb 25.8 m ↗	↓	^{210}Pb 22.3 y ↗	↓	^{206}Pb stable		^{211}Pb 36.1 m ↗	↓ ^{207}Pb stable
Tl			^{210}Tl 1.3 m		^{206}Tl 4.2 m			^{207}Tl 4.79 m ↗	

↓ alpha decay; ↗ beta decay; half life (d = days; m = minutes; s = seconds; y = years)

Source: NCRP 1975

$$\left(\frac{^{206}Pb}{^{204}Pb}\right) = \left(\frac{^{206}Pb}{^{204}Pb}\right)_t + \left(\frac{^{238}U}{^{204}Pb}\right)[\exp(\lambda^{238}t) - 1]$$

The first term on the right side refers to the radiogenic isotope ratio at any time in the past, t, in the history of a natural object including that at its formation. In the latter case, it is known as the initial ratio. Note that the last three equations refer to particular cases of the general equation, Equation (5.19). Numerous applications of these decay systems in different geological contexts can be found in the works of Faure and Mensing (2005), Dickin (2005), Allegre (2008), and

DePaolo (1988). A few prominent applications of these systems will be given in later chapters in appropriate contexts.

Table 5.1 shows one non-radiogenic isotope (d') for each of the daughter elements that is conventionally used in age equations. If the daughter element contains more than one non-radiogenic isotope, the ratios of these isotopes in any natural sample will be the same today as at any time in the past. Thus, any measured deviation of these ratios from their accepted or universal values can be used to correct the measured and desired radiogenic isotope ratio for any analytical aberrations, as will be explained in the next chapter on analytical methods. Argon, calcium, strontium, hafnium, osmium, and neodymium have two or more non-radiogenic isotopes to provide this analytical advantage. Unfortunately, Pb has only one non-radiogenic isotope (^{204}Pb) precluding a very precise measurement of its three radiogenic isotope ratios.

The decay constants of parent isotopes in Table 5.1 were calculated mainly from their disintegration rates [Equation (5.1)] as measured by counting the α, β, or gamma (γ) emissions from a precisely known quantity of the isotope. The low disintegration rates of very long-lived parents [Equation (5.1)] make their absolute measurement very difficult. Hence, the experimental uncertainties in most constants are still large, with some in the percent level. Villa and Renne (2005) have summarized the values and uncertainties of decay constants of the parent isotopes in Table 5.1. High accuracy and precision of decay constant values are necessary only when ages measured by two or more decay pairs on the same sample are to be tightly compared. Uncertainties in a decay constant do not matter much for tight comparisons of ages measured using a single parent-daughter system.

5.4 Short-lived Parent-daughter Couples used in Radiometric Dating

Table 5.3 lists another group of parent-daughter systems, but with half-lives much shorter than the age of the Solar System. Any of these short-lived isotopes that was present in the freshly isolated Solar System material 4.5 billion years ago must have decayed to below detection long time ago. While this is true for some nuclides, like ^{244}Pu (of plutonium) and ^{146}Sm, quite a few are present now at measurable levels. The presence of ^{234}U, ^{230}Th, ^{226}Ra, and ^{210}Po can be readily understood, in the light of the foregoing discussion, as due to their continuous production from the decay of their ultimate long-lived parent, ^{238}U (Table 5.2). The present-day presence of the first few parent isotopes is due to their generation by natural nuclear reactions (Chapter 2) caused by cosmic rays on the abundant target nuclei in the atmosphere, and the top silicate layers of the earth.

Protons and α particles in the galactic cosmic rays are so energetic that they cause high energy nuclear reactions in target nuclides when they strike the top of the earth's atmosphere, shattering or 'spalling' them into stable and unstable nuclides of smaller mass and atomic number, and releasing cascades of secondary particles including neutrons, protons, and mu-mesons. These secondary particles cause additional, but low energy reactions in the lower atmosphere, resulting in the production of stable and unstable isotopes of many elements. A small fraction of the secondary radiation actually reaches the surface of the earth to produce new nuclides in the exposed soil and rock (Lal, 1988). The cosmic-ray-produced isotopes are collectively known as cosmogenic isotopes, although all the isotopes of chemical elements in the Solar System are ultimately cosmogenic in the sense of their genesis in cosmic settings. The most important cosmogenic radionuclides are

^{3}He (of helium), ^{10}Be (of beryllium), ^{14}C (of carbon), ^{26}Al (of aluminium), ^{32}Si (of silicon), ^{36}Cl (of chlorine), ^{39}Ar, and ^{81}Kr (of krypton). Several of these nuclides have sufficiently long half-lives to be useful for tracking or tracing recent geologic processes. We will restrict ourselves to just two prominent cosmogenic isotopes, ^{14}C and ^{10}Be, produced by the interaction of secondary and primary cosmic rays, respectively, with abundant nuclei in the atmosphere.

Table 5.3 Parent-daughter Systems with Half-lives Comparable or Less than 10^8 Years

p'	p	d	d'	Mode	$\lambda\ (y^{-1})$	Half-life (–)
^{9}Be	^{10}Be	^{10}B	^{11}B	β^-	4.6×10^{-8}	1.5×10^7
^{12}C	^{14}C	^{14}N	^{15}N	β^-	1.21×10^{-4}	5730
^{27}Al	^{26}Al	^{26}Mg	^{24}Mg	β^-	9.7×10^{-7}	715×10^5
^{35}Cl	^{36}Cl	^{36}Ar	^{40}Ar	β^+	2.25×10^{-6}	3.08×10^5
^{40}Ca	^{41}Ca	^{41}K	^{40}K	β^+	6.7×10	1.0×10^5
^{55}Mn	^{53}Mn	^{53}Cr	^{56}Cr	β^-	1.89×10^{-7}	3.39×10^6
^{56}Fe	^{60}Fe	^{60}Ni	^{58}Ni	β^-	4.76×10^{-7}	1.46×10^6
^{106}Pd	^{107}Pd	^{107}Ag	^{109}Ag	β^-	1.06×10^{-7}	6.4×10^6
^{128}I	^{129}I	^{129}Xe	^{128}Xe	β^-	4.0×10^{-8}	17.3×10^6
^{144}Sm	^{146}Sm	^{142}Nd	^{144}Nd	α	6.85×10^{-9}	101×10^6
^{181}Hf	^{182}Hf	^{182}W	^{183}W	β	7.69×10^{-8}	9.01×10^6
–	^{210}Po	^{206}Pb	^{204}Pb	α	3.11×10^{-2}	22.26
–	^{226}Ra	^{222}Rn	–	α	4.27×10^{-4}	1.62×10^3
–	^{230}Th	^{226}Ra	–	α	9.22×10^{-6}	755×10^4
–	^{234}U	^{230}Th	^{232}Th	α	2.79×10^{-6}	25×10^5
–	^{244}Pu	$^{131-135}$Xe	^{128}Xe	sf		8.4×10^6

Note: sf stands for spontaneous fission.

Low energy neutrons from secondary cosmic rays react with the most abundant ^{14}N (of nitrogen) atoms in the atmosphere to produce radioactive ^{14}C according to:

$$^{14}N + n \rightarrow {}^{14}C + p$$

^{14}C beta-decays back again to ^{14}N with a half-life of about 5,700 years to reach an equilibrium concentration between decay and growth. This ^{14}C quickly oxidizes to $^{14}CO_2$ (of carbon dioxide), mixes with the far more abundant stable $^{12}CO_2$ and $^{13}CO_2$ molecules, and is globally homogenized by atmospheric circulation before being absorbed photosynthetically by plants or exchanged with CO_2 in water. The former eventually enters all living organisms through the food chain, and the latter may be deposited as carbonates. As long as an organism is alive and exchanging CO_2 (directly or indirectly) with the atmosphere, its tissues will have the equilibrium atmospheric ^{14}C concentration. Once the exchange of the organism with the atmosphere stops due to its death or other reasons, the equilibrium ^{14}C abundance begins to decrease exponentially with time according to the basic radioactive decay equation [Equation 5.8)], which, in this specific case, becomes:

$$(^{14}C) = (^{14}C)_{eq}\exp(-\lambda t) \qquad (5.22)$$

Note that t is now the time since isolation of a living organism from the atmosphere, and (^{14}C) is the residual content of ^{14}C. If the cosmic-ray production of ^{14}C in the atmosphere at the time of death of organism is assumed to be the same as it is now, $(^{14}C)_{eq}$ will be as measured in any living organism. This relation between absolute concentrations of ^{14}C in the beginning and at present can be changed to an equivalent relation between their disintegration rates [activity according to Equation (5.5)] by multiplying both sides by λ, as:

$$(\lambda^{14}C) = (\lambda^{14}C)_{eq}\exp(-\lambda t) \qquad (5.23)$$

The activity $(\lambda^{14}C)_{eq}$ determined by counting β emissions is 13.6 disintegrations per minute (dpm) per gram of living or modern C, and is called the specific activity. An organic C sample with exactly half of this specific activity must have ceased living exactly one half-life time ago; therefore, its radiocarbon age is 5,700 years. The practical upper limit to radiocarbon dating based on disintegration rates of necessarily very large quantities (a few grams) of C is about 35,000 years. The idea of using radiocarbon as a dating tool, particularly in archeology, was conceived and implemented by W F Libby, for which he received the Nobel Prize in chemistry in 1960. (Arnold and Libby, 1949).

The number of ^{14}C atoms in a gram of modern C can be calculated by substituting its specific activity in Equation 5.5 after dividing the decay constant, $1.209 \times 10^{-4}\, y^{-1}$ (Table 5.1) by the number of minutes ($365 \times 24 \times 60$) in a year to convert it to its equivalent value in inverse minutes:

$$13.6 = (^{14}C)_{eq} \times \left[\frac{1.209 \times 10^{-4}}{(365 \times 24 \times 60)}\right]$$

The number of atoms in $(^{14}C)_{eq}$ is huge, 5.8×10^{10}. Even after 10 half-lives (57,000 years) the sample will still contain about 58 million surviving atoms, suggesting that a substantial increase in the range and a decrease in sample size for radiocarbon dating are possible if the residual atoms can be directly determined. The relevant age equation is:

$$\left(\frac{^{14}C}{^{12}C}\right) = \left(\frac{^{14}C}{^{12}C}\right)_{eq} \exp(-\lambda t) \qquad (5.24)$$

As the number of stable ^{12}C isotopes in 1 gram of C is 5.0×10^{22}, the ratio on the right side is extremely small, that is, $\sim 10^{-12}$. The capability to measure such low C isotope ratios was realized only in the 1970s through a clever integration of a conventional low voltage mass spectrometer and a high voltage (a few million volts) particle accelerator, aptly called the Accelerator Mass Spectrometer (AMS). The AMS, briefly described in the next chapter, has become the preferred method for the detection of ^{14}C and many other cosmogenic isotopes, because it is much faster and requires much less sample quantities than that based on decay counting. The reader is referred to Faure and Mensing (2005) and Dickin (2005) for many applications of radiocarbon analysis in earth and archeological sciences using both methods. One unusual and striking application of AMS is, however, worth mentioning here. A piece of cotton cloth, called the 'Shroud of Turin', was believed to have been wrapped over the crucified body of Christ. However, its sudden appearance only in the 1350s made its age crucial to its authenticity. Radiocarbon dating of this cloth by β counting was ruled out, as it would have destroyed much of this venerable relic. But, simultaneous

AMS dating in three labs required a total only about 150 mg even before removing its very likely younger C contamination from human handling. The results showed convincingly that the cotton used in the cloth grew in AD 1290 ± 25 year, postdating Christ by about seven centuries (Damon et al., 1989).

In practice, radiocarbon dating has to contend with several complications like variation in cosmic-ray production of ^{14}C with time, its dilution by the addition of dead C (older than 1,00,000 years) in the atmosphere from the large scale burning of fossil fuels after the industrial revolution, and the artificial production of ^{14}C in the atmosphere by atmospheric nuclear bomb tests in the 1950s (Faure and Mensing, 2005; Dickin, 2005).

The highly energetic protons and α particles in the primary cosmic rays also produce much lighter, stable, and radioactive nuclides through their interactions at the top of the atmosphere. ^{7}Be with a half-life of 53 days and ^{10}Be with half-life of 1.51 My represent two such spallation-produced nuclides, ^{10}Be being lighter than its target isotopes ^{14}N and ^{16}O by four and six nucleons, respectively. ^{10}Be is now measured relative to its only stable counterpart, ^{9}Be on an AMS. Unlike C, Be does not form a gas and is scavenged from the atmosphere by rain before it can be homogenized globally. The flux of ^{10}Be on the surface of the earth is, therefore, far from uniform. However, in the oceans the distribution of ^{10}Be is uniform because it resides long enough in ionic form or in suspension to be mixed well. ^{10}Be is extracted from seawater by adsorption on suspended clay particles, which sink to be part of a growing sediment column. This forms the basis for the measurement of rates of sediment accumulation or of accretion of manganese (Mn) nodules in oceanography. The principle behind the two applications is as follows. ^{10}Be in each layer of a sediment column will decay exponentially with time, t as:

$$(^{10}Be) = (^{10}Be)_t \exp(-\lambda t)$$

AMS measurement directly of $^{10}Be/^{9}Be$ ratio also avoids measurement of absolute concentrations of ^{10}Be.

$$\left(\frac{^{10}Be}{^{9}Be}\right) = \left(\frac{^{10}Be}{^{9}Be}\right)_t \exp(-\lambda t) \qquad (5.25)$$

If the layer is now at a depth h in the sediment column, and the sedimentation rate is uniform at a, t in the equation can be replaced by (h/a).

$$\left(\frac{^{10}Be}{^{9}Be}\right) = \left(\frac{^{10}Be}{^{9}Be}\right)_t \exp\left[-\lambda\left(\frac{h}{a}\right)\right]$$

Taking natural logarithm,

$$\ln\left(\frac{^{10}Be}{^{9}Be}\right) = \ln\left(\frac{^{10}Be}{^{9}Be}\right)_t - \left(\frac{\lambda}{a}\right)h \qquad (5.26)$$

This means that a plot of the left hand side versus depth for sediment samples taken from known depths below the sediment-water interface in a given location (as in a sediment core) will be a straight line, provided the ^{10}Be/^{9}Be ratio at the time of deposition of the sampled layers has remained practically the same over time. The slope of this line, (λ/a), will be a measure of the sedimentation rate. The procedure is essentially the same to measure rate of growth of Mn nodules, which are small, rounded, and concentrically-banded concretions found on the deep ocean floor containing abundant manganese oxide.

Many other applications of ^{10}Be, as well as examples for many other chemically different radioactive nuclides such as ^{26}Al, ^{32}Si, ^{36}Cl, ^{39}Ar, ^{41}Ca, ^{53}Mn, and ^{81}Kr can be found in Faure and Mensing (2005) and Dickin (2005). One elegant application of ^{10}Be is, however, worth mentioning here, as it is also imaginatively creative. It has to do with the recycling (subduction) of surface sediments into the magma source of island arc volcanics, as envisaged in the revolutionary theory of Plate Tectonics (Chapter 10).

For our present purpose, it is sufficient to think of an island arc as a linear or arcuate chain of volcanic islands (e.g. Japanese or Aleutian islands in the Pacific Ocean) situated on the ocean floor and parallel to a deep trench (Figure 10.8). The trench marks the zone where a rigid segment of the earth's surface or plate plunges, or is subducted at an angle (the so called Benioff zone) below the island chain. This subduction causes the melting of the overlying material at some depth and the melt rises to the surface to form a chain of volcanoes overhead. The conspicuous presence of cosmogenic ^{10}Be in deep sea sediments (~10^9 atoms/gram) and its absence deep in the earth provide an elegant way to assess any possible entrainment or assimilation of subducted ocean-floor sediment into the island arc magma source. A positive ^{10}Be signal in arc volcanics would imply not only sediment assimilation, but also that the transport time from sedimentation to the magma source region is less than a few half-lives of ^{10}Be. In a first study, Brown et al. (1982) found a significant excess of ^{10}Be in the island arc materials relative to other volcanic rocks on the ocean-floor or continents, and attributed it to the involvement of subducted ocean-floor sediment in the genesis of island-arc volcanics. Subsequent studies showed that not all arc volcanics have a positive signal. However, a component derived from the subducting slab has now been inferred from many other evidences.

It was mentioned above that a small fraction of the secondary cosmic rays actually reaches the surface of the earth to produce new nuclides in the exposed soil and rock; these are known as in situ cosmogenic isotopes. Cosmic rays are so much attenuated by the earth's atmosphere that in situ production on exposed rocks was not even looked for initially. Nevertheless, they were looked for much earlier on atmosphere-free solid objects, like meteorites and lunar rocks exposed directly to primary cosmic rays. It is only with the development of AMS that the feeble generation of cosmogenic isotopes in the thin silicate layers on the solid earth could be detected and quantitatively measured. Lal (1988) gave a comprehensive review of in situ cosmogenic isotopes and scientific applications, thereof.

It should be noted that radiometric dating using short-lived nuclides differs from that using long-lived nuclides. The former requires measurement of only the parent isotope, whereas, the latter requires both parent and its daughter nuclides. Strictly speaking, the former belongs to the category of radioactive isotope geochronometers, and the latter to radiogenic isotope geochronometers. The term 'radiometric dating' covers both. Less obvious is the fact that parent nuclide concentrations in the former case are invariably very small even in the beginning of the decay interval.

5.5 INTERPRETATION OF *t*, RADIOACTIVE DECAY INTERVAL

When any of the generalized equations developed in Section 5.2 are applied to the practical problem of determining the age of actual samples, attention must be given to the question of the meaning of the term age, which has been used as a synonym for the time interval, *t*, in the foregoing equations. Specifically, *t* represents the interval of time between the present time, and the time in the past when the number of atoms of the daughter isotope was d_t. For *t* to represent this time interval correctly, the method requires that the number of atoms of both the parent and daughter elements have changed only as a consequence of radioactive decay of the parent, as well as the daughter isotope, if it is also radioactive. The system is then said to be closed with respect to parent and daughter. It is essential in every case to be able to characterize the time of some significant event in the history of the sample with a correct value of the appropriate quantity, p_t or d_t. Growth in a closed system is called 'single stage growth'. Strictly speaking, single stage growth, or evolution means that (p/d') and (d/d') ratios in the system changed only due to radioactive decay.

We will now consider how the fundamental age equation is used to calculate the unknown variable, *t*, or 'age', as is usually meant.

$$\left(\frac{d}{d'}\right) = \left(\frac{d}{d'}\right)_t + \left(\frac{p}{d'}\right)[\exp(\lambda_p t) - 1]$$

This can be written compactly as:

$$\alpha = \alpha_t + \mu[\exp(\lambda_p t) - 1], \tag{5.27}$$

remembering that quantities without a right suffix represent present-day values that, at least in principle, can be measured in the laboratory. $\alpha = (d/d')$ is the present day isotopic ratio of the daughter element, $\alpha_t = (d/d')_t$ that ratio at a time *t* in the past, and $\mu = (p/d')$ is the parent-daughter composition ratio.

When $t \ll (1/\lambda)$, Equation (5.21) simplifies to the linear relation:

$$\alpha = \alpha_t + \mu \lambda_p t \tag{5.28}$$

Except in the case when α_t is zero or negligible, its knowledge is required to calculate the other variable, *t*. However, if α_t is known to vary within small limits, and $\ll \alpha$, *t* can be calculated from a single sample. Such ages are called model ages. Model ages can be calculated in quite different ways (Chapter 11), having the only the common property that calculation of an age value requires some additional assumption(s).

5.6 ISOCHRON CONCEPT, ISOTOPE EQUILIBRATION, AND CLOSURE TEMPERATURE

The change from α_t to α in an object depends on the decay interval *t* and the relative proportions of parent and daughter elements, expressed as $\mu = (p/d')$. The dependence of α on *t* and μ can be separated mathematically by taking the partial derivative of α with respect to μ and *t* (Bevington, 1969). Let us first consider the effect of varying μ in a set of samples having the same age, *t*, and initial ratio, α_t. The partial derivative of α with respect to μ, $(\partial \alpha / \partial \mu)$ is:

$$\left(\frac{\partial \alpha}{\partial \mu}\right) = [(\exp(\lambda t) - 1]$$ (5.29)

This shows that if α is plotted as ordinate against μ as abscissa for this set of samples, the data will define a straight line with a slope, $[\exp(\lambda t) - 1]$, as shown in Figure 5.3.

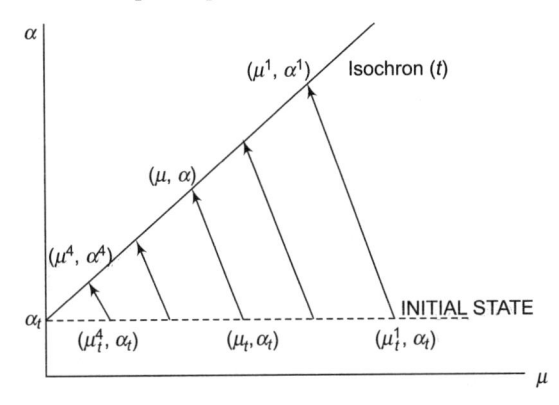

Figure 5.3 Isotope evolution diagram showing the correlation of $(d/d' = \alpha)$ with $(p/d' = \mu)$ for a long-lived nuclide, p. The slope of this line (isochron) gives the 'age' and its intersection at $(p/d') = 0$, the initial state. (Gopalan, 2015)

The meaning of this kind of diagram is that the slope of the line determines the time at which all members had the same α_t given by the y-intercept. This line is called an isochron—the loci of points of the same age. The isochron diagram was conceived first in 1961 by L O Nicolaysen of the Bernard Price Institute of Geophysical Research, University of Witwatersrand, South Africa (Nicolaysen, 1961). It is now a widely used geochronological tool, applicable in one form or another, to all of the decay schemes used for radiometric dating. While the isochron is linear in this plot, it may not be so in various types of diagrams or plots used in radiogenic isotope geochemistry. The most straightforward use of the age equation and diagram is to calculate the time elapsed since a set of systems with varying μ possessed, or were made to possess a common starting α_t. The mechanism for achieving a uniform initial isotopic composition could result from a variety of processes in nature like magmatic melting, solid state diffusion (as during metamorphic heating), dissolution in an aqueous medium, and mixing in the primordial solar nebula. The equation simply represents a mathematical relationship; its particular meaning requires additional information, especially the process which equilibrated the isotopes in the systems on the scale of their sampling.

In practice, the measured points representing the different systems may fit closely, but not exactly on a straight line (York, 1966). The deviation of the measured points from a perfect straight line arises in two ways:

1. Experimental error in both the measured ratios, α and μ.
2. Small, but significant differences in t and α_t among the systems and/or failure of the closed system condition.

One can estimate the experimental error associated with the individual measured points and apply statistical tests to determine if the deviations from a straight line can be explained entirely on the basis of such errors (Chapter 7). The standard criterion for this is that a statistical parameter, called

Mean Square Weighted Deviates (MSWD), is about unity (York, 1966). Then the line is called an isochron—locus of points of the same age and initial daughter composition. A MSWD greater than 1 is interpreted to mean that the deviations exceed the limits of analytical error and must be due to failure in different degrees of the set of systems to conform to isochron conditions. Such poorly-fit straight lines are sometimes called errorchrons (McIntyre et al., 1966; Brooks et al., 1972). This term must be used with great discretion, as a best fit straight line will be a valid isochron if the experimental errors are large, but will become an errorchron if the analytical errors are reduced.

A linear array of several systems does not necessarily have a time connotation. For example, variable mixtures of two unrelated or arbitrary components with different (p/d') and (d/d') will also form a linear array, but without any time significance. This linear correlation results from the denominators, d' being the same for the ratios plotted on both ordinate, and abscissa (Chapter 4 in this book; Langmuir et al., 1977; Faure and Mensing, 2005). A linear correlation with a negative slope can be easily recognized as a mixing line. But a line with a positive slope could be mistaken for an isochron, and, hence, must be regarded with caution.

Two important aspects of isotope geochemistry emerge from the forgoing discussion of an isochron and a mixing line. If a chemical system is fractionated or differentiated into two more subsystems (ss) with different μ, the daughter isotope composition immediately after fractionation in each subsystem will be the same. They will become isotopically distinct only with the passage of time. However, when two isotopically distinct systems are mixed, the mixture will differ from both right at the time of mixing. The phrases 'possessed' or 'made to possess' a common isotopic composition refer, respectively, to the situation just before the differentiation of a single homogeneous chemical system, and after isotopic equilibration of an initially inhomogeneous chemical system.

Isotopic equilibration or homogenization mentioned in the foregoing refers to the equalization or elimination of any isotopic differences within, and between different phases of a geochemical system of arbitrary size. It is brought about by the net transport of atoms relative to their immediate surroundings by their random movements, and is called diffusion. Diffusion is characterized by a parameter called the diffusion coefficient, D. Various types of atomic or ionic diffusion in solids, liquids, and gases, and their role in the kinetics of melting, crystallization of magmatic systems, and recrystallization in metamorphic systems are covered in many text books (Philpotts, 1990). Our interest is mainly the range of distances over which isotopic equilibration can be realized by diffusion alone. The average diffusion distance, x, covered by a diffusing species in a given medium is given approximately as (Allegre, 2008):

$$x = (D\ t)^{0.5}$$

where, x is in centimetres, t in seconds, and D in $cm^2 s^{-1}$. In rocks at room temperatures, $D \sim 2 \times 10^{-36}$ $cm^2 s^{-1}$ (Allegre, 2008), which gives x as less than a millimetre even after $t \sim 10^{36}$ years. This shows that even microscopic mineral grains will retain their isotopic integrity quantitatively for billions of years, a strict precondition for reliable radiometric dating. Diffusion coefficients of liquids and gases are several orders of magnitude greater than those in solids even at room temperatures, leading to correspondingly large diffusion distances in them. Even so, diffusion alone cannot equilibrate isotopes on a regional scale, even less so on planetary and nebular scales. Such a large scale isotope homogenization in liquids and gases is, however, possible because of a very much faster bulk transport process, called advection (Philpotts, 1990; Albarede, 2003). Viewed in this context, the quotation in the beginning of this article by O'Keefe conveys in an intuitively suggestive way the enormous difference in diffusion ranges over geologically long times in the three states of matter. A physically trapped microscopic pocket of liquid and gas will, nonetheless, retain its isotopic identity or integrity for geologically long periods. Well known examples are tiny fluid inclusions, or gas-filled vesicles in many minerals.

We next consider the extremely sensitive dependence of the diffusion coefficient and, hence, diffusion distances of an atom on temperature. This dependence is given by:

$$D = D_0 \exp\left(-\frac{E}{RT}\right)$$

where, D_0 (cm²/s) is called the 'frequency factor', E (cal/mol) the activation energy imposed by a crystalline medium on the diffusing species, R (1.985 cal/°C/mol) the gas constant, and T the absolute temperature. The spectacular changes in D even over a small change in temperature (100°C) are given by York and Farquhar (1972) in the case of Ar diffusing through a mineral. For a typical E of 40 kcal/mol, D is $10^{-29} D_0$ at $T = 27°C$ (300 K) and $2.10 \times 10^{-22} D_0$ at $T = 127°C$ (400K), a change in D by a factor of $\sim 2 \times 10^7$ with a corresponding change in diffusion distances of Ar atoms in a mineral in a given time. This implies that the transition of a mineral grain from being wide open to diffusive loss of Ar (diffusion distance comparable to grain size) to its quantitative retention (diffusion distance much smaller than grain size) could occur within a very narrow temperature range. The temperature below which Ar retention is quantitative is called the blocking, or more appropriately, closure temperature (Dodson, 1973), and marks the onset of Ar accumulation and, hence, the turning on of the radioactive clock.

D depends sensitively also on D_0 and E which are characteristic of the diffusing species and the medium through which it diffuses. This means that different minerals will close at different temperatures and different isotopic systems in the same mineral will also close at different temperatures. In other words, different minerals and different parent-daughter systems in a slowly cooling geologic body will register systematically varying ages. If the closure temperatures are known, they can be combined with the measured ages to construct a cooling curve for the host rock .This field is called Thermochronology (McDougall and Harrison, 1999).

5.7 Termination of a Single Stage Growth in the Past

If the evolution from time t in the past is terminated at a later time, t_1, but before the present, by the withdrawal of the parent element, the present day composition will be the difference in evolution over time t, and the evolution over the shorter interval t_1 with a present parent-daughter ratio, μ.

$$\alpha = \alpha_t + \mu[\exp(\lambda t) - 1] - \mu[\exp(\lambda t_1) - 1] = \alpha_t + \mu[\exp(\lambda t) - \exp(\lambda t_1)] \quad (5.30)$$

It must be noted that equations (5.30) and (5.13) are equivalent.

5.7.1 Multi-stage stage evolution

The single stage growth arrested at t_1 shows how to calculate the present isotopic composition of a system which evolved from t to t_1 in one reservoir with μ_1, and from t_1 to t_2 ($t_2 \neq 0$) in a second reservoir with μ_2, μ_1, and μ_2 being as measured at present:

$$\alpha = \alpha_t + \mu_1[\exp(\lambda t) - \exp(\lambda t_1)] + \mu_2[\exp(\lambda t_1) - \exp(\lambda t_2)].$$

If the second stage ends at the present time, $t_2 = 0$:

$$\alpha = \alpha_t + \mu_1[\exp(\lambda t) - \exp(\lambda t_1)] + \mu_2[\exp(\lambda t_1) - 1]. \qquad (5.31)$$

5.7.2 n-stage evolution

The two stage evolution is readily extended to any number of stages (say 5) of evolution as:

$$\alpha = \alpha_{tp} + \mu_1[\exp(\lambda t_p) - \exp(\lambda t_1)] + \mu_2[\exp(\lambda t_1) - \exp(\lambda t_2)] + \mu_3[\exp(\lambda t_2) - \exp(\lambda t_3)]$$
$$+ \mu_4[\exp(\lambda t_3) - \exp(\lambda t_4)] + \mu_5[\exp(\lambda t_4) - 1] \qquad (5.32)$$

where, μ for each stage is calculated as it would be measured today, t_p, the time in the past marking the beginning of the first stage. This equation is for n stages involving $(n - 1)$ fractionations in separate reservoirs. The evolution of the isotopic composition of an element through five stages of differentiation is shown in Figure 5.4a. Figure 5.4b shows the evolution of five stages of differentiation each followed by instantaneous mixing (as shown by a, b, c, d) with an isotopically distinct component(s). It is easy to appreciate that Figure 5.4b will become a continuous curve if a chemical system suffers no fractionation, but only thorough mixing of a single element with distinctly different isotopic compositions from two or more sources. The most striking example of this case is the time variation of Sr isotope composition in sea water (Veizer, 1989), which is the basis of Sr isotope stratigraphy (Banner, 2004).

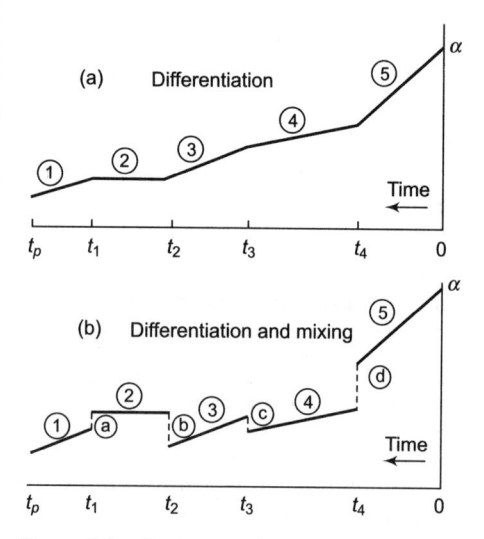

Figure 5.4 Five stage evolution of a chemical system with only differentiation at the beginning of each stage (a), and with both differentiation and mixing (b). (Gopalan, 2015)

Equation (5.27) for single stage evolution and Equation (5.32) for multistage (five stage) evolution are mathematically similar. However, they do not have the same significance. In single stage evolution, α and μ can, in principle, be measured in a sample to calculate a unique value for t, if α_{tp} is known. Whether t is a geologically meaningful age depends on the satisfaction of conditions discussed already. There is no such unique solution to the multi-stage Equation (5.32), as values of $\mu_1, \mu_2, \mu_3,$ and μ_4, and $t_1, t_2, t_3,$ and t_4 are generally unknown, or unknowable. Even in the simple case of a two stage evolution [Equation (5.31)], one must make educated guesses about μ_1 and t_1 for the evolution of α_{tp} to α (Stacey and Kramers, 1975). However, the multistage equation can be used to deduce the time-average value of the μ's in the five stages. Using, for simplicity, the linear approximation of the exponential function, Equation (5.32) can be rewritten as:

$$\alpha - \alpha_{tp} = \lambda\mu_1(t_p - t_1) + \lambda\mu_2(t_1 - t_2) + \lambda\mu_3(t_2 - t_3) + \lambda\mu_4(t_3 - t_4) + \lambda\mu_5(t_4 - 1) = \lambda\mu_a t_p$$

μ_a is the time average of the parent-daughter ratios in the five stages, and represents the effective parent-daughter ratio in an equivalent single stage evolution over the entire duration, t_p. A single stage evolution has an explicit geochronological significance, whereas, multistage evolution has an implicit geochemical significance. The latter potential in the case of the decay of ^{87}Rb to ^{87}Sr was first pointed out by Paul Gast in his pioneering paper (Gast, 1960) in the following words:

The isotopic composition of a particular sample of strontium, whose history may or may not be known, may be the result of times spent in a number of such systems or environments. In any case the isotopic composition is the time-integrated result of the Rb/Sr ratios in all the past environments.

5.8 RECOGNITION OF EVENTS CAUSING ISOTOPIC EQUILIBRATION ON DIFFERENT SCALES

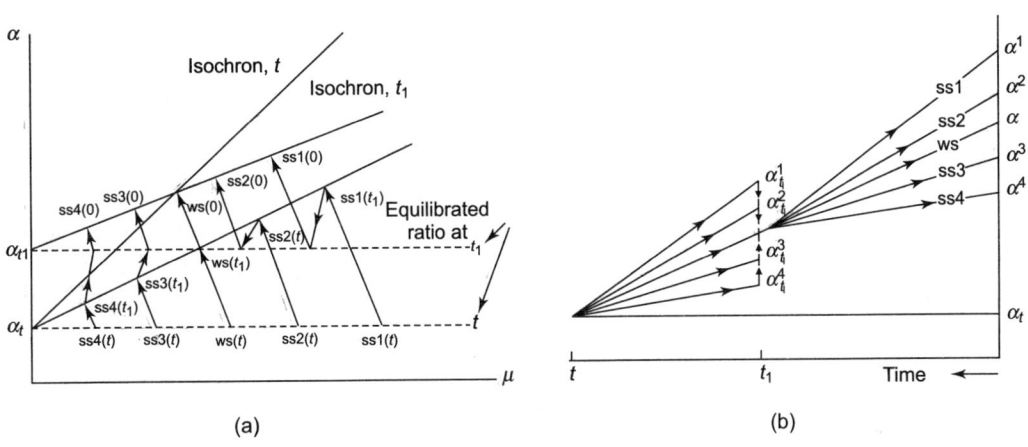

Figure 5.5 Parent-daughter evolution on (α, μ) (a) and (α, t) (b) diagrams showing four ss initially synchronous with their parent source, which were internally rehomogenized later. (Gopalan, 2015)

Consider now that a natural object (ws) with a homogeneous composition (α_t, μ_t) underwent a chemical fractionation to form (say) four subsystems (ss) at time t in the past. Assume that two subsystems had their μ (μ_t^1, μ_t^2) increased, and the other two subsystems had lowered μ (μ_t^3, μ_t^4) relative to the whole system (ws) (Figure 5.5a). All subsystems will, however, have the same isotope composition (α_t) as the ws. From the mixing theory (Chapter 4 in this book; Faure and Mensing, 2005):

$$\mu_t = f^1\mu_t^1 + f^2\mu_t^2 + f^3\mu_t^3 + f^4\mu_t^4 \tag{5.33}$$

where, f^1, f^2, f^3, and f^4 are the factors by which the composition ratios were changed. The isotopic evolution from α_t at t to a later time t_1 in the ws, and the four subsystems (Figure 5.5b):

ws(t_1) $\qquad \alpha_{t1} = \alpha_t + \mu_t[1 - \exp - \lambda(t - t_1)]$ $\hfill$ (5.34)

ss1(t_1) $\qquad \alpha^1_{t1} = \alpha_t + \mu_t^1[1 - \exp - \lambda(t - t_1)]$

ss2(t_1) $\qquad \alpha^2_{t1} = \alpha_t + \mu_t^2[1 - \exp - \lambda(t - t_1)]$

ss3(t_1) $\qquad \alpha^3_{t1} = \alpha_t + \mu_t^3[1 - \exp - \lambda(t - t_1)]$

ss4(t_1) $\qquad \alpha^4_{t1} = \alpha_t + \mu_t^4[1 - \exp - \lambda(t - t_1)]$

Multiplying the second of the above equations by f_1, third by f_2, etc., and summing, gives:

$$f^1\alpha^1_{t1} + f^2\alpha^2_{t1} + f^3\alpha^3_{t1} + f^4\alpha^4_{t1} = \alpha_t(f^1+f^2+f^3+f^4) + (f^1\mu_t^1+f^2\mu_t^2+f^3\mu_t^3+f^4\mu_t^4)[1-\exp-\lambda(t-t_1)]$$
$$= \alpha_t + \mu_t[1-\exp\lambda(t-t_1)] = \alpha_{t1}$$

Thus, the weighted mean isotope composition of the mixture of the four subsystems at t_1 is exactly equal to that of the ws at the same time (Figure 5.5b). If only the subsystems were isotopically equilibrated at time t_1, but the whole rock remained chemically closed, subsequent evolution to the present will be:

ws(0)	$\alpha = \alpha_t + \mu_t[1-\exp-(\lambda t)]$	(5.35)
ss1(0)	$\alpha^1 = \alpha_{t1} + \mu_t^1\exp-\lambda(t-t_1)[1-\exp-(\lambda t_1)]$	(5.36)
ss2(0)	$\alpha^2 = \alpha_{t1} + \mu_t^2\exp-\lambda(t-t_1)[1-\exp-(\lambda t_1)]$	
ss3(0)	$\alpha^3 = \alpha_{t1} + \mu_t^3\exp-\lambda(t-t_1)[1-\exp-(\lambda t_1)]$	
ss4(0)	$\alpha^4 = \alpha_{t1} + \mu_t^4\exp-\lambda(t-t_1)[1-\exp-(\lambda t_1)]$	

This shows that the ws represents a single stage evolution over a longer interval, t, whereas, the four subsystems each represents a single stage evolution over a smaller interval, t_1, but from an evolved common composition, α_{t1}. The four subsystems will define an isochron with slope $[\exp(\lambda t_1) - 1]$, and intercept α_{t1}, while the ws together with other ws (not shown) will define an isochron with slope $[\exp(\lambda t) - 1]$, and intercept at α. This means that isotopic equilibration on a mineral scale during metamorphism and that on a whole-rock scale during magmatism can facilitate dating both events (Fairbairn, 1961; Lanphere et al., 1964). This element transport takes place even when textural and petrographic manifestation of a thermal episode is not evident in the rocks (Wasserburg, 1987).

5.9 Projection of Present Day Isotopic Composition Back in Time

The standard form of the fundamental age equation gives the present day isotopic composition that evolved from a former composition in a system with a given present day parent-daughter composition, μ. The terms of this equation can be transposed to calculate former isotopic compositions for given present day values, namely α and μ, as:

$$\alpha_t = \alpha - \mu[\exp(\lambda t) - 1] \tag{5.37}$$

α_t in this equation is now a variable, and must not be confused with a fixed initial daughter composition, as had been considered so far. This is a mathematically trivial operation. However, this form of the age equation has an historical importance, and has more recently given insights into large scale geodynamical processes in a way not obvious from the isochron approach. If $t \ll (1/\lambda)$, this equation approximates to:

$$\alpha_t = \alpha - \mu\lambda t$$

The partial derivative of α_t with respect to time, keeping α and μ constant, gives the rate of change of isotopic composition in the past:

$$\left(\frac{\partial \alpha_t}{\partial t}\right) = -\mu\lambda$$

This means that the present day α will, with time, decrease into the past linearly with a slope that is proportional to μ for a given radioactive species. The trajectories of the daughter ratios in two arbitrary systems will, therefore, meet at some arbitrary time in the past, as two lines with unequal slopes will eventually intersect. But, a set of systems which define an isochron on a plot of α vs μ (Figure 5.3) will all converge at a single point corresponding to their common initial ratio and t in a plot of α vs t, as shown in Figure 5.6. It is obvious that if the present day α and μ for all the four systems are known, their common age will correspond to the time when all had the same daughter isotope ratio.

It must be noted at this stage that both (α, μ) and (α, t) diagrams are silent on the parent material of the chemical systems plotted on them. Figure 5.7 shows that the parent material can be just one or many. The three chemical systems evolving from t, in either diagram, are shown to be similar for simplicity, but they need not be so. Figure 5.7a shows that a single homogeneous chemical system, s, evolves until t, when a part of it fractionates into two subsystems, d^1 and d^2. The unfractionated part continues its evolution into t space. Figure 5.7b, on the other hand, shows that two or more chemical systems, s^1 and s^2 evolve independently until time t, when both merge in appropriate proportions into a chemically and isotopically homogeneous body, mimicking the chemistry of the single source in the other case. This body subsequently fractionates into three subsystems, m^1, m^2, and m^3. The important point to note is that d^1 and d^2 will inherit the isotope composition of their parental source at time t, whereas, m^1, m^2, and m^3 will differ isotopically from either s^1 or s^2, and have no memory of their antecedent history. This simple diagram depicts the two important processes, namely differentiation and mixing, controlling the chemical geodynamics of the earth, in particular, and planets, in general (Allegre, 1987). Acting alone, differentiation would have created by now a hopelessly diverse suite of chemical systems in and on the earth. Mixing alone would have produced a homogeneous, and, hence, uninteresting and featureless earth. It is the interplay over geologic time of these two opposing processes of

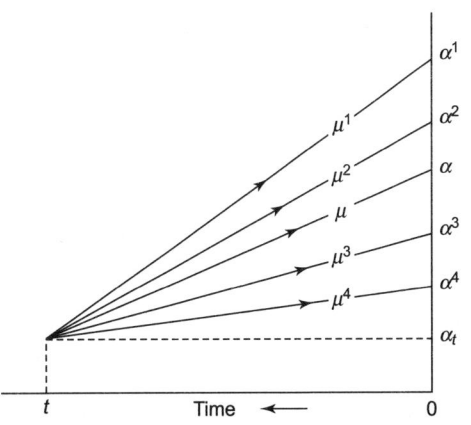

Figure 5.6 Projection of five parent-daughter systems back in time when they shared the same daughter isotopic composition. (Gopalan, 2015)

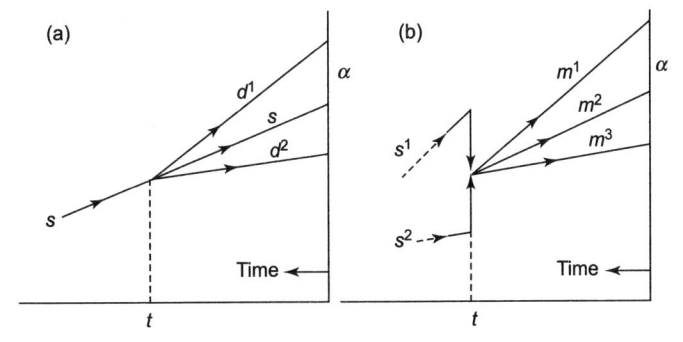

Figure 5.7 Three derivatives from a single parent source (a) and from a homogenized mixture of two parent sources (b). (Gopalan, 2015)

divergence and convergence that has led to the presently observable mosaic of chemical systems. Both processes need energy to drive them.

5.10 Reservoir Ages

The differentiation of an originally uniform and large reservoir is the basis of the so called 'reservoir model' of age calculation. The concept of reservoir model is shown in Figure 5.8. In this diagram, the line marked 'r' depicts the single stage evolution or trajectory of a well-mixed reservoir large enough not to be affected by the chemical extraction of small subsystems at different times from it. The two branching lines labelled s and s^1, respectively, represent two subsystems derived by chemical fractionation of the reservoir at two different times. If the reservoir parameters are labelled 'r' and parameters of the subsystems are labelled s, the time of extraction of the latter will be when both had the same daughter isotope ratio. That is:

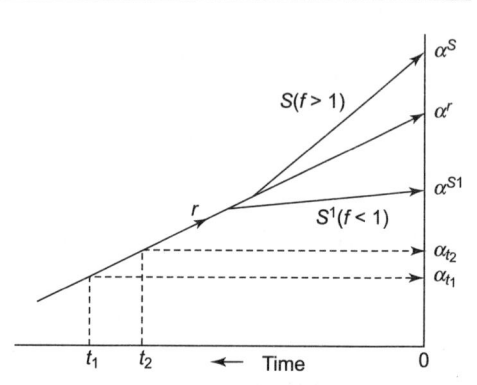

Figure 5.8 Evolution of derivatives (s and s_1) from a well-mixed large and known reservoir, r. (Gopalan, 2015)

$$\alpha^t - \mu^r[\exp(\lambda t) - 1] = \alpha^s - \mu^s[\exp(\lambda t) - 1], \qquad (5.38)$$

Or,
$$t = \left(\frac{1}{\lambda}\right) \ln\left[1 + \frac{(\alpha^s - \alpha^r)}{(\mu^s - \mu^r)}\right]$$

The t so calculated is a model age, as it is based on the assumption of the isotopic composition of the reservoir being copied into the subsystems at the time of its extraction. It should be noted that reservoir ages are, in fact, two-point isochron ages.

Differential comparison of a sample isotopic ratio relative to that of a reference standard is a common practice in stable isotope geochemistry of H, C, and oxygen(O) since it was introduced around 1950 in Harold Urey's laboratory in the University of Chicago. DePaolo and Wasserburg (1976) have adapted this technique in radiogenic isotope geochemistry specifically for differential comparison of the radiogenic isotope ratio of Nd in a sample relative to that in a large universal reservoir, called the Chondritic Uniform Reservoir (CHUR) with a chondritic (meteoritic) Sm/Nd ratio. The time since a sample separated from a reservoir of chondritic composition is called a Chondritic Model age (Chapter 11). Model ages can also be calculated to represent the time of separation of a sample from other plausible types of reservoirs (Rollinson, 1993). As deviations of Nd isotope ratios relative to CHUR are in parts per 10,000, they introduced an epsilon notation as:

$$\varepsilon_t = \left[\left(\frac{\alpha^s}{\alpha^r}\right) - 1\right] \times 10{,}000$$

where, t refers to the time in the past corresponding to α^s and α^r, and the superscripts s and r to sample and CHUR, respectively. Since differences in Nd ratio result from differences in parent-daughter ratios, μ^s and μ^r, DePaolo and Wasserburg (1976) also defined a fractionation factor for concentration ratios as:

$$f = \left[\left(\frac{\mu^s}{\mu^r}\right) - 1\right]$$

If $f > 0$, then the system is said to be enriched in the parent element relative to the reservoir, and if $f < 0$, it is said to be depleted. The two cases are illustrated in Figure 5.8.

An interesting case is one in which two subsystems separated from the reservoir at different time t_1 and t_2 (as shown in the lower part of the figure) with negligible or zero μ as measured today. Then:

$$\alpha_{t1} = \alpha^r - \mu^r[\exp(\lambda t_1) - 1]$$
$$\alpha_{t2} = \alpha^r - \mu^r[\exp(\lambda t_2) - 1],$$

where, $t_1 > t_2$

$$(\alpha_{t2} - \alpha_{t1}) = \mu^r[\exp(\lambda t_1) - \exp(\lambda t_2)] \tag{5.39}$$

The shift in α can, therefore, be related to passage of time if the reservoir parameters are known or assumed (Wasserburg, 1987).

For $t > 1/\lambda$, the linear approximation is not valid. The basic nonlinear equation has to be used to trace the past variation of daughter isotope composition as:

$$\alpha_t = \alpha - \mu[\exp(\lambda t) - 1] \tag{5.40}$$

Values of time t, measured positively into the past, have been plotted in figures 5.4 to 5.8, and also in later figures, on the negative side of the x-axis. Some authors (Wasserburg, 1987; DePaolo, 1988) plot these on the positive side of the x-axis. The latter will be images of the former in a plane mirror along the y-axis. The arrows point in the direction of the passage of time.

5.11 Coupling Two Different, But Chemically Identical Parent-Daughter Systems

We have so far seen how the data on a single parent-daughter pair in several chemical systems can be combined in order to check their closed system evolution, and then calculate their common age and initial ratio. If the test fails, one can try some other parent-daughter system which may be geochemically more robust or resistant. Simultaneous use of two decay systems has been found to be far more effective than their separate or sequential uses. The age equations for two different parent-daughter pairs, in general, are:

$$\alpha^1 = \alpha_t^1 + \mu^1[\exp(\lambda_1 t) - 1] \tag{5.41}$$
$$\alpha^2 = \alpha_t^2 + \mu^2[\exp(\lambda_2 t) - 1] \tag{5.42}$$

where, $\alpha^1, \alpha^2, \mu^1$, and μ^2 refer to the two decay systems with decay constants, λ_1 and λ_2, respectively. Although even two chemically different decay pairs can, in principle, be coupled, we will first consider coupling two chemically identical decay systems, as they are simpler to interpret. It so happens that the only known chemically identical decay pairs are ^{238}U-^{206}Pb and ^{235}U-^{207}Pb systems (Table 5.2) that were extensively and exclusively studied during the pioneering work on geochronology in the early part of the last century. Thus, until recently, simultaneous use of coupled parent-daughter systems has been restricted to these two U-Pb pairs. These two pairs have been coupled in two different ways to eliminate the need for the measurement of parent isotopes, or to correct for open system effects in U-rich minerals. It is instructive to study these two approaches in a general way, but in the light of studies of the two U-Pb decay pairs (^{238}U-^{206}Pb, and ^{235}U-^{207}Pb, respectively).

Consider a parent element consisting of two long-lived radioactive isotopes (p^1 and p^2) decaying to two different isotopes (d^1 and d^2) of a daughter element with decay constants, λ_1 and λ_2, respectively. Their equations in the explicit form are:

$$\left(\frac{d^1}{d'}\right) = \left(\frac{d^1}{d'}\right)_t + \left(\frac{p^1}{d'}\right)[\exp(\lambda_1 t) - 1] \quad (5.43)$$

$$\left(\frac{d^2}{d'}\right) = \left(\frac{d^2}{d'}\right)_t + \left(\frac{p^2}{d'}\right)[\exp(\lambda_2 t) - 1] \quad (5.44)$$

where, d' is a non-radiogenic isotope of the daughter element. The change in the daughter isotope compositions from time t in the past to the present is shown individually in Figure 5.9 for the case of $\lambda_2 > \lambda_1$.

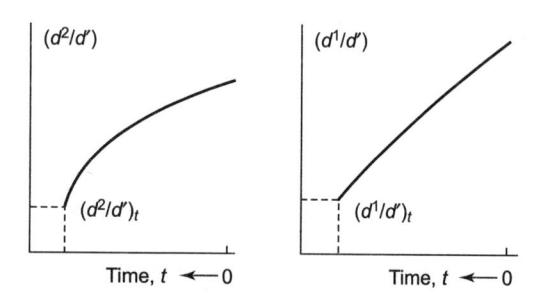

Figure 5.9 Evolution of two independent parent-daughter systems with different decay constants. (Gopalan, 2015)

For $1/\lambda_2 < t < 1/\lambda_1$, both curves are noticeably nonlinear. But the curve for λ_2 increases rapidly in the beginning and flattens out later, while the curve for λ_1 shows a more uniform rate of change. These two diagrams can be combined into a single diagram (Figure 5.10) by plotting (d^2/d') vs (d^1/d') for various values of t. The curve IJKL, commonly called a growth curve, is the locus of the two daughter isotope ratios in a closed reservoir. In other words, it represents the single stage evolution of two chemically identical parent-daughter systems. Time is represented on this curve by a series of points. For example, daughter isotope compositions were at I at the very beginning t years ago, at J t_1 years ago, at K t_2 years ago, and are at L at present.

By transferring the first term on the right-hand side to the left side, and dividing Equation (5.44) by the other, we obtain:

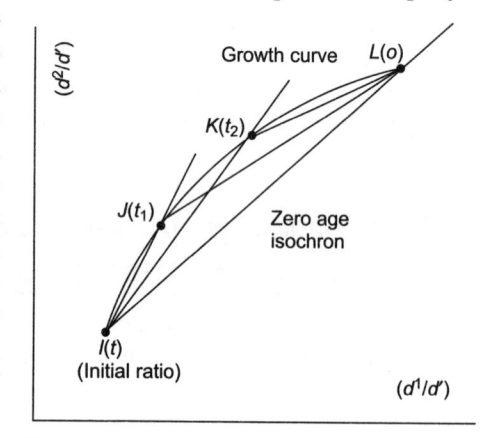

Figure 5.10 Growth curve for two chemically identical parent-daughter systems. (Gopalan, 2015)

$$\frac{\left[\left(\frac{d^2}{d'}\right) - \left(\frac{d^2}{d'}\right)_t\right]}{\left[\left(\frac{d^1}{d'}\right) - \left(\frac{d^1}{d'}\right)_t\right]} = \frac{\left(\frac{p^2}{d'}\right)}{\left(\frac{p^1}{d'}\right)} \cdot \frac{[\exp(\lambda_2 t) - 1]}{[\exp(\lambda_1 t) - 1]}$$

This can be written in a compact form as:

$$\frac{[\alpha^2 - \alpha_t^2]}{[\alpha^1 - \alpha_t^1]} = \left(\frac{\mu^2}{\mu^1}\right)\frac{[\exp(\lambda_2 t) - 1]}{[\exp(\lambda_1 t) - 1]} \qquad (5.45)$$

The important feature of this equation is that it needs only isotope ratios of the parent and daughter elements, and not their absolute concentrations, unlike in the case of the individual equations. In fact, only the daughter isotope ratios are required, as the present parent isotope ratio is a universal constant and known independently. So, if the initial daughter isotope compositions are known for a single sample, its age can be calculated from the measured gross values. This equation is, however, transcendental, and so the right-hand side cannot be solved algebraically for t, which is usually solved iteratively (Faure and Mensing, 2005.).

Equation (5.45) is of the form $(y - y_0)/(x - x_0)$ = constant for a given t. This is the general equation of a straight line passing through the point (x_0, y_0) with a slope equal to the constant. Thus, Equation (5.45) is that of the straight line IL. It is an isochron, as all systems with different μ values, but evolving from the same initial composition, I and all sampled at the present time will have the same slope. Time t can be calculated by measuring the slope of this isochron, setting it equal to the right-hand side and solving the resulting transcendental equation. But, unlike the isochrons based on a single parent-daughter pair, the initial ratio I cannot be determined by its intercept on either coordinates of this graph. Line IL can also result from mixtures of two end member components with composition I and L, respectively. This is a general property of mixing lines on plots of the form (a/b) vs (c/d), where $b = d$, which is the case here with $d' = b = d$ (Chapter 4 of this book; Langmuir et al., 1977; Faure and Mensing, 2005).

It can be shown that any other line segment in this diagram (like JL, KL, IJ, and IK) is also an isochron. For example, consider the main system at a particular time t_1 in the past corresponding to Point J in the diagram. Its parent-daughter ratios will be:

$$(\mu^1)_J = (\mu^1)\exp(\lambda_1 t_1)$$
$$(\mu^2)_J = (\mu^2)\exp(\lambda_2 t_1)$$

where $(\mu^1) = (p^1/d')$ and $(\mu^2) = (p^2/d')$ are the present day ratios. Let this system undergo geochemical fractionation such that some portions of the system have their μ increased and this ratio decreased in other portions. The two parent-daughter ratios in each of the subsystems can be represented as:

$$(\mu^1)_{nJ} = f_n(\mu^1)_J; \quad (\mu^2)_{nJ} = f_n(\mu^2)_J$$

where, f_n is the factor by which this ratio changed in the subsystems n at time J, as a result of the geochemical fractionation. Because parent and daughter isotopes are isotopes of the same elements, f_n is the same for both decay systems. For each ss, the present-day isotopic compositions of its daughter will be given by:

$$(\alpha^1)_n = (\alpha^1)_{nJ} + f_n(\mu^1)\exp(\lambda_1 t_1)\exp(1 - \exp(-\lambda_1 t_1)]$$
$$= (\alpha^1)_{nJ} + f_n(\mu^1)[\exp(\lambda_1 t_1) - 1]]$$
$$(\alpha^2)_n = (\alpha^2)_{nJ} + f_n(\mu^2)[\exp(\lambda_2 t_1) - 1]$$

The present-day daughter isotope compositions of the subsystems will all plot along a straight line passing through J and L. This can be seen immediately by setting $f_n = 0$ and $f_n = 1$. The fact that the line is straight can be shown by differentiating both equations with respect to the parameter f_n and taking the ratio of the two derivatives:

$$\frac{d(\alpha^2)_n}{d(\alpha^1)_n} = \left(\frac{\mu^2}{\mu^1}\right)\frac{[\exp(\lambda_2 t_1)-1]}{[\exp(\lambda_1 t_1)-1]}$$

For a given value of t_1, the right side of equation is a constant. Thus, the locus of the points $[(\alpha^1)_n, (\alpha^2)_n]$ has constant slope, that is, the line is straight. This line is an isochron because it is the locus of points, all of which experienced geochemical fractionation at time t_1 to form subsystems with differing parent-daughter ratios. Subsystems for which $f_n < 1$ would plot between J and L; those with $f_n > 1$ would plot on the continuation of JL beyond Point L. The time in the past at which the main system was chemically fractionated can be calculated from the slope of JL.

The age equations for the line segments JL, KL, IJ, and IK are:

$$\text{JL,} \quad \frac{\left[\left(\frac{d^2}{d'}\right)-\left(\frac{d^2}{d'}\right)\right]_{t1}}{\left[\left(\frac{d^1}{d'}\right)-\left(\frac{d^1}{d'}\right)\right]_{t1}} = \left(\frac{p^2}{p^1}\right)\frac{[\exp(\lambda_2 t_1)-1]}{[\exp(\lambda_1 t_1)-1]} \qquad (5.46)$$

$$\text{KL,} \quad \frac{\left[\left(\frac{d^2}{d'}\right)-\left(\frac{d^2}{d'}\right)\right]_{t2}}{\left[\left(\frac{d^1}{d'}\right)-\left(\frac{d^1}{d'}\right)\right]_{t2}} = \left(\frac{p^2}{p^1}\right)\frac{[\exp(\lambda_2 t_2)-1]}{[\exp(\lambda_1 t_2)-1]} \qquad (5.47)$$

$$\text{IJ,} \quad \frac{\left[\left(\frac{d^2}{d'}\right)-\left(\frac{d^2}{d'}\right)\right]_{t}}{\left[\left(\frac{d^1}{d'}\right)-\left(\frac{d^1}{d'}\right)\right]_{t}} = \left(\frac{p^2}{p^1}\right)\frac{[\exp(\lambda_2 t)-\exp(\lambda_2 t)]}{[\exp(\lambda_1 t)-\exp(\lambda_1 t_1)]} \qquad (5.48)$$

$$\text{IK,} \quad \frac{\left[\left(\frac{d^2}{d'}\right)-\left(\frac{d^2}{d'}\right)\right]_{t}}{\left[\left(\frac{d^1}{d'}\right)-\left(\frac{d^1}{d'}\right)\right]_{t}} = \left(\frac{p^2}{p^1}\right)\frac{[\exp(\lambda_2 t)-\exp(\lambda_2 t_2)]}{[\exp(\lambda_1 t)-\exp(\lambda_1 t_2)]} \qquad (5.49)$$

These four straight lines are also isochrons. JL and KL represent the respective times t_1 and t_2 in the past when the reservoir fractionated with initial ratios corresponding to J and K. IJ and IK represent the duration of closed system evolution of the reservoir until t_1 and t_2 years ago, respectively. IL represents the closed evolution of the reservoir until the present time, and hence is called the zero age isochron. In the case of the two U-Pb pairs, the zero age isochron is called the geochron.

Equation (5.45), as applied to the U decay systems ^{238}U-^{206}Pb and ^{235}U-^{207}Pb is the basis of the so-called Pb-Pb age, introduced by Nier (1939) in his classic and pioneering discussion of U-lead dating of U minerals. The straight line, defined by this equation in a plot of (^{207}Pb/^{204}Pb) vs (^{206}Pb/^{204}Pb) was first termed isochrone by Houtermans. Equations (5.48) and (5.49), as applied to U-Pb system, represent what today is known as the Holmes-Houtermans U-Pb model to calculate the 'age of the Earth' (Holmes, 1946; Houtermans, 1946). The dependence of age equations solely on daughter isotope ratios facilitates calculation of reliable ages that are not possible from the individual decay equations, for example, in the case of very recent loss or gain of parent element. It must be noted that the slope is only weakly time dependent, if the decay constants are very close to each other.

5.12 Concordia and Discordia

A Pb-Pb isochron, in particular, and a daughter-daughter isochron, in general, based on the above coupling of two chemically coherent decay systems require that samples used to define it evolved in a chemically closed condition. Loss or gain of parent and daughter elements during the course of their individual evolution due to open system processes will result in an unacceptable scatter of data. The present understanding of geochemical processes causing partial loss or gain of the two elements is not sufficiently quantitative to allow correction. But, if two decay pairs are chemically coherent, the loss or gain of parent and/or daughter will tend to affect or change both parent-daughter ratios by the same factor. This feature is utilized in a different way of coupling the two decay schemes to identify open system effects, and in favourable cases, even to apply a correction. This correction procedure was a great advance in geochronology inasmuch as open system effects are the rule rather than the exception, especially in decay systems with a large chemical contrast between the parent and daughter elements, as in the case of U-Pb pairs. This coupling scheme was, in fact, first conceived and implemented by G W Wetherill (1956) in the case of U minerals.

The terms of the age equation can be transposed to give the ratio of the radiogenic daughter to the parent isotope in one decay scheme as:

$$\frac{\left[\left(\dfrac{d^1}{d'}\right)-\left(\dfrac{d^1}{d'}\right)_t\right]}{\left(\dfrac{p^1}{d'}\right)} = \exp(\lambda_1 t) - 1$$

$$\left(\frac{d^{1*}}{p^1}\right) = \exp(\lambda_1 t) - 1 \tag{5.50}$$

where d^{1*} is the radiogenic addition over time t for a present-day, $\mu = (p^1/d')$. A similar equation for the second decay pair is:

$$\left(\frac{d^{2*}}{p^2}\right) = \exp(\lambda_2 t) - 1 \tag{5.51}$$

The two ratios on the left-hand side of equations (5.50) and (5.51) are different exponential functions of the same variable, t. The values of the two functions can be calculated for different values of t and plotted, as shown in Figure 5.11. The two functions in the case of the U-Pb decay pairs will be (^{207}Pb*/^{235}U) and (^{206}Pb*/^{238}U), respectively. Figure 5.11 for U-Pb pairs was called the Concordia diagram by its inventor, G W Wetherill. Concodia is the locus of concordant ages, that is, the two ages calculated from the two equations are the same. Discordant ages will obviously fall away from this curve. Though this curve looks similar to the curve in the previous daughter-daughter plot (d^2/d') vs (d^1/d') (Figure 5.10), it is different from the latter in many respects. In the Concordia plot, (1) the coordinates are different; (2) the scale of the two axes is invariant; (3) the curve begins at the origin corresponding to zero age; (4) the ages increase along the curve; and (5) the line joining any two points on this curve is not a line of equal ages (isochron), but, as we will see soon, unequal or discordant ages.

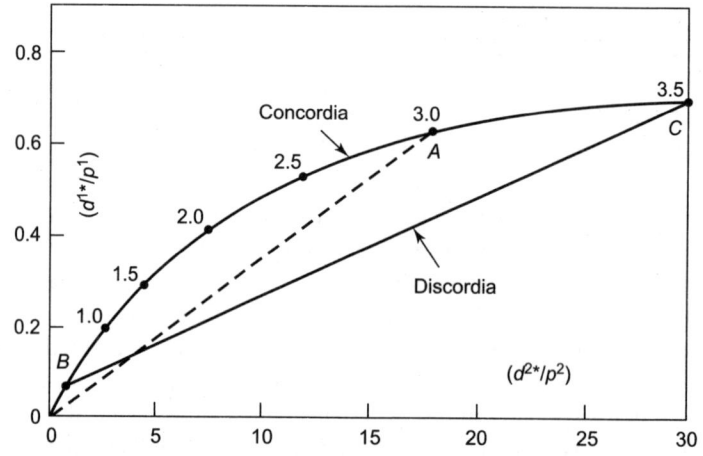

Figure 5.11 Concordia-discordia plot. The x- and y-axes will be (^{207}Pb*/^{235}U) and (^{206}Pb*/^{238}U), respectively, in this plot for the U-Pb decay pairs (Wetherill, 1956). (Gopalan, 2015)

The use of the Concordia diagram (Figure 5.11) can be visualized by considering the hypothetical history of a set of minerals which evolved in a closed condition for, say, 3,000 million years in the past, and underwent partial loss of daughter element at the present time due to an event, such as brief metamorphic heating. Starting at the origin 3,000 million years ago, the grains would move together along the Concordia curve, reaching Point A corresponding to 3,000 million years of age. Although the mineral grains will lose their daughter atoms to different extents, the two daughter-parent ratios in each grain will be reduced by the same factor, f, because of their chemical coherence. Grains losing their entire daughter will be shifted to the origin, whereas, those with different degrees of daughter loss will lie on the line connecting Point A with the origin, O. As the points on this line represent a set of paired discordant ages, it is called a Discordia. If the grains become chemically closed once again and evolve for say, 500 million years into the future, grains originally at the origin will move to Point B corresponding to 500 million years of age, and grains at Point A with zero loss ($f = 1$) would move to Point C corresponding to an age of 3,500 million years. The grains on the Discordia will also evolve, but remain discordant even after 500 million years on a new Discordia, BC.

Wetherill (1956) derived the time dependence of parent-daughter ratios in the two U-Pb decay pairs due to episodic, but partial daughter loss (Pb). This case of abrupt midcourse change in the parent-daughter ratio is mathematically equivalent to the daughter evolution in two reservoirs with different μ's. The general equation, Equation (5.25) for two stage evolution is:

$$\left(\frac{d}{d'}\right) = \left(\frac{d}{d'}\right)_t + \mu_1[\exp(\lambda t) - \exp(\lambda t_1)] + \mu_2[\exp(\lambda t_1) - 1]$$

where, μ_1 and μ_2 refer to the two reservoirs, as measured today. In the present case, μ_2 is the measurable parameter in the second reservoir, and $\mu_1 = f \cdot \mu_2$, where f is the factor by which μ_1 in the first reservoir was shifted abruptly to μ_2 at time t_1. Substituting for μ_1, and rearranging the terms as in Equation (5.43), we get:

$$\left(\frac{d^*}{p}\right) = f[\exp(\lambda t) - \exp(\lambda t_1)] + [\exp(\lambda t_1) - 1] \tag{5.52}$$

The two decay pairs will be distinguished as:

$$\left(\frac{d^{1*}}{p^1}\right) = f[\exp(\lambda_1 t) - \exp(\lambda_1 t_1)] + [\exp(\lambda_1 t_1) - 1] \tag{5.53}$$

$$\left(\frac{d^{2*}}{p^2}\right) = f[\exp(\lambda_2 t) - \exp(\lambda_2 t_1)] + [\exp(\lambda_2 t_1) - 1] \tag{5.54}$$

Because both parent and daughter isotopes are isotopes of the same elements, f is the same for both decay schemes. These equations are valid not only for loss of the daughter element but also for the gain/loss of the parent element. However, gain of the extraneous daughter with a different isotopic composition will break the continuity of daughter isotope evolution.

The present day parent-daughter ratios in mineral grains, as calculated from equations (5.53) and (5.54) will plot on a straight line passing through points B and C. That B and C will fall exactly on the Concordia can be readily checked by setting $f = 0$ and $f = 1$ corresponding to total loss and no loss, respectively, of the daughter element. That this straight line will also include grains with intermediate f values can be shown by differentiating both equations with respect to the parameter, f, and taking the ratio of the two derivatives:

$$\frac{d\left(\dfrac{d^{1*}}{p^1}\right)}{d\left(\dfrac{d^{2*}}{p^2}\right)} = \frac{\left[\exp(\lambda_1 t) - \exp(\lambda_1 t_1)\right]}{\exp(\lambda_2 t) - \exp(\lambda_2 t_1)} \tag{5.55}$$

For given values of t and t_1, the right side is a constant. Thus, the locus of intermediate points has a constant slope and hence is a straight line.

The parameter f in equations (5.53) and (5.54) defines the magnitude of the open-system effects. As this parameter approaches unity, the corresponding mineral grains approach closed system evolution, and, therefore, provide a set of paired discordant ages converging toward their common, true concordant age. In other words, the upper intercept of the Discordia BC with the Concordia defines the time of original crystallization ($f = 1$), whereas, the lower intercept with Concordia ($f = 0$) gives the time of episodic open-system process. Although episodic, but partial loss of daughter element is the most common type of open-system behavior, other and very complex causes of such behavior are conceivable (Tilton, 1960; Wasserburg, 1963).

Equations (5.50) and (5.51) give two other exponential functions of the single variable t as:

$$\left(\frac{d^{2*}}{d^{1*}}\right) = \left(\frac{p^2}{p^1}\right)\left[\frac{(e^{\lambda_2 t} - 1)}{(e^{\lambda_1 t} - 1)}\right] \quad (5.56)$$

$$\left(\frac{p^1}{d^{1*}}\right) = \frac{1}{[e^{\lambda_1 t} - 1]} \quad (5.57)$$

The plot of (d^{2*}/d^{1*}) vs (p^1/d^{1*}) for different values of t defines another Concordia curve. This curve for the two U-Pb decay pairs corresponds to plotting $(^{207}\text{Pb}/^{206}\text{Pb})^*$ against $(^{238}\text{U}/^{206}\text{Pb}^*)$. This was proposed by Tera and Wasserburg (1974) to interpret U-Th-Pb data from lunar rocks, and is called the Tera-Wasserburg (or inverse) Concordia diagram. This version of the Concordia diagram has some advantage over the conventional Wetherill Concordia curve, in that the most accurately determined ratio, $^{207}\text{Pb}/^{206}\text{Pb}$, is plotted directly, permitting more immediate understanding of the extent to which the experimental data constrain their interpretation. This plot has also been preferred in dating very young rocks because it displays Discordia lines more clearly than the conventional diagram (Dickin, 2005).

5.13 Coupling Two Chemically Different Decay Systems

We will consider the specific case of coupling the two chemically different decay pairs, ^{87}Rb-^{87}Sr and ^{147}Sm-^{143}Nd, as it has provided valuable insights into the mantle melting processes (Chapter 11). Equations (5.41) and (5.42) can now be used to represent Rb-Sr and Sm-Nd systems, respectively:

$$\alpha^1 = \alpha_t^1 + \mu^1[\exp(\lambda_1 t) - 1] \quad (5.58)$$

$$\alpha^2 = \alpha_t^2 + \mu^2[\exp(\lambda_2 t) - 1] \quad (5.59)$$

where, $\alpha^1 = (^{87}\text{Sr}/^{86}\text{Sr})$, $\mu^1 = (^{87}\text{Rb}/^{86}\text{Sr})$, $\alpha^2 = (^{143}\text{Nd}/^{144}\text{Nd})$, and $\mu^2 = (^{147}\text{Sm}/^{144}\text{Nd})$ are the present ratios. As the half-lives of ^{87}Rb and ^{147}Sm are much larger than the age of the earth, the above equations can be simplified to linear form as:

$$\alpha^1 = \alpha_t^1 + \mu^1 \lambda_1 t \quad (5.60)$$

$$\alpha^2 = \alpha_t^2 + \mu^2 \lambda_2 t \quad (5.61)$$

Dividing Equation 5.61 by Equation 5.60 gives:

$$\frac{(\alpha^2 - \alpha_t^2)}{(\alpha^1 - \alpha_t^1)} = \left(\frac{\mu^2}{\mu^1}\right)\left(\frac{\lambda_2}{\lambda_1}\right) = \text{constant} \quad (5.62)$$

This is the equation of a straight line with the constant as its slope. This line IJKL in Figure 5.12 is analogous to the growth curve IJKL in Figure 5.10, and is the locus of Sr and Nd isotope compositions developed in a uniform reservoir with present day μ^1 and μ^2. Consider the system at a later time t_1 in the past, corresponding to Point J in the figure. The parent-daughter ratios at this time will be:

$$(\mu^1)_J = \mu^1 \exp(\lambda_1 t_1)$$

$$(\mu^2)_J = \mu^2 \exp(\lambda_2 t_1)$$

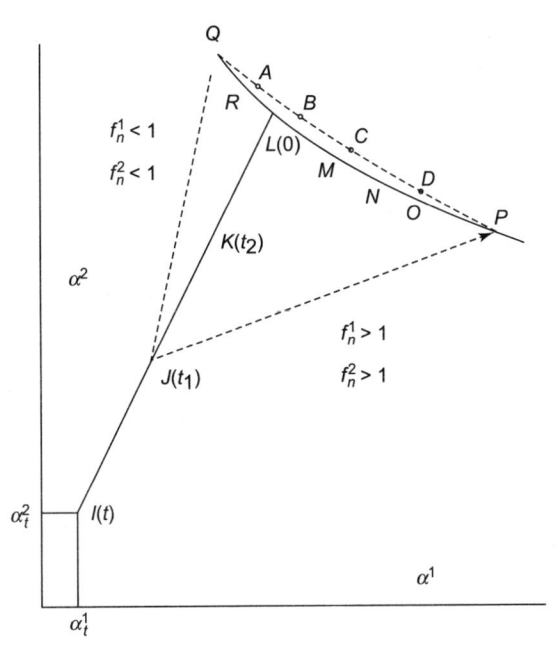

Figure 5.12 Coupling of two chemically different parent-daughter pairs. (Gopalan, 2015)

Now, suppose the system at J is geochemically fractionated into n subsystems with their parent-daughter ratios changed from that of the ws by different factors, f_n^1 and f_n^2. The parent-daughter ratios of each of the subsystems at time t_1 will then be described by the expressions:

$$(\mu^1)_{nJ} = f_n^1(\mu^1)_J; \quad (\mu^2)_{nJ} = f_n^2(\mu^2)_J$$

f_n^1 and f_n^2 will, in general, be unequal, unlike in the case of the two chemically identical U-Pb pairs. For each of the ss, the present-day Sr and Nd isotope compositions will be given by:

$$(\alpha^1)_{nL} = (\alpha^1)_J + f_n^1(\mu^1)\lambda_1 t_1 \tag{5.63}$$

$$(\alpha^2)_{nL} = (\alpha^2)_J + f_n^2(\mu^2)\lambda_2 t_1 \tag{5.64}$$

Dividing one by the other gives:

$$\frac{[(\alpha^2)_{nL} - (\alpha^2)_J]}{[(\alpha^1)_{nL} - (\alpha^1)_J]} = \frac{(f_n^2)}{(f_n^1)}\left(\frac{\mu^2}{\mu^1}\right)\left(\frac{\lambda_2}{\lambda_1}\right)$$

A subsystems with (Rb/Sr) ratio higher than in the ws will have (Sm/Nd) less than that in the ws because of the differential extraction of Rb, Sr, Sm, and Nd during melting. In other words, if f_n^1 is > 1, f_n^2 will be < 1. So, this subsystems will evolve along a line JP with a smaller slope than that of JL representing the original system. Points O, N, and M represent the present-day Sr and Nd isotope compositions of other subsystems formed at time t_1, but with different f_n^1 and f_n^2. Points Q and R on the other side of the growth curve IL represent the isotopic compositions of subsystems with f_n^1 < 1 and f_n^2 > 1. The curve defined by points P, O, N, M, L, R, and Q is the locus of subsystems formed at the same time from the same initial Sr and Nd ratios, and hence is an isochron. In this case, the isochron is, in general, a curve instead of a straight line because, unlike the U-Pb systems, neither the numerators nor the denominators of the ratios plotted on the ordinate and abscissa are isotopes of the same element. If the origin is shifted to Point L, the isochron would extend to the left of the origin into the second quadrant and to its right into the fourth quadrant. This diagram forms the basis of tracking the fractionation of the bulk silicate earth or primitive mantle into an upper mantle depleted in incompatible elements and continental crust enriched in the same elements (Chapter 11).

The curve QABCDP is the locus of arbitrary mixtures of the two extreme components P and Q. It will, in general, be a hyperbola (Chapter 4) with its curvature decided by the ratio $(^{144}Nd/^{86}Sr)_Q$ to $(^{144}Nd/^{86}Sr)_P$. If this ratio is < 1, the curve will be concave upward and convex upward when this ratio is > 1. Only when this ratio is exactly 1 will this mixing line will be a straight line. While both isochrons and mixing lines are curves in this case, the latter need not go through the origin, and, hence, a part of it may lie either in the first quadrant or in the third.

5.14 Chemical and Half-life Diversity of Parent-daughter Pairs

We recall from earlier sections that a radiometric age represents the time of the latest fractionation of a parent element from its daughter element during natural geological processes. If the fractionation is quantitative, the daughter nuclide measured in a sample is entirely radiogenic. If the fractionation is only partial, as in most cases, one has to allow for the initial presence of the daughter element. So we will consider the chemical contrast between the parent and daughter in each pair and the chemical diversity of the individual pairs listed in Table 5.1. In the K-Ar system, the parent is a chemically active alkali element and the daughter is a chemically inert gas. They are, therefore, readily and quantitatively fractionated. In the K-Ca and Rb-Sr pairs, the parents are two alkali elements and the daughters are two alkaline earth elements. The chemical contrast within each pair is sufficient for their efficient fractionation in most cases. However, the K-Ca system is not useful as a geochronometer, as the decay of a trace isotope (0.0119 per cent) of potassium (^{40}K) into a dominant isotope (98%) of Ca (^{40}Ca) produces barely measurable radiogenic Ca, except in special cases (Marshall and DePaolo, 1982). Sm and Nd are both rare earth elements with very similar chemical affinities, and, hence, are not easily fractionated. This resistance to fractionation is a distinct advantage in some petrogenetic contexts (Wasserburg, 1987). Rhenium and osmium belong to the so-called geochemical family of siderophile elements (Chapter 10), but can be quite fractionated during geological processes. However, their extremely low abundance in crustal and mantle rocks presents an analytical challenge. The chemical contrast between U and Pb is quite strong. In addition, the two U isotopes (Table 5.1) decay to two separate Pb isotopes with very different half-lives. These features have been innovatively exploited in both Pb-Pb and U-Pb dating, as described earlier.

Wasserburg (1987) has succinctly summarized the choice of parent-daughter systems as follows:

The various nuclear chronometers and their host phases have different reaction rates and susceptibility to thermal history. Those which are most sensitive or compliant, record the time of the most recent event to which they are susceptible. Those chronometers and phases which are more robust will remember earlier stages of development. So in order to understand the phenomenon under investigation, one should choose a chronometric system that registers the process that controls the phenomenon. There are different clocks and rock forming-reforming processes. There are no bad chronometers, only bad interpretations of them.

5.15 Radiometric Dating by Indirect Radiogenic Effects

We noted in the beginning that the decay of a radioactive parent nuclide results in the formation of a daughter nuclide (slightly lighter than its parent), a few much lighter particles (neutrinos, betas and alphas), and release of decay energy (typically from a few tens of KeV to a few MeV). Conservation of linear momentum requires that the light particles carry the bulk of the decay energy as their kinetic energy. The daughter nuclide with very small kinetic energy comes to rest very close to the site of its decayed parent. With time, the accumulation of such nuclides forms the basis of radiogenic isotope geochronology, as elaborated so far. Of the light 'radiogenic particles', neutrinos interact so tenuously with matter, as to escape with their energy intact. Beta and α particles dissipate their kinetic energy by ionizing and exciting atoms over a small range in the crystal before coming to rest. Betas lose their identity after slowing down, whereas, the alphas retain their identity as He atoms. The accumulation of He atoms can be used as the accumulation of the heavier daughter nuclides for geochronological purposes. In fact, Rutherford used the production of He from the decay of U for the very first geochronological determination. Decay of U and Th to Pb isotopes was realized much later.

The ionization of atoms in non-conducting crystalline solids by the passage of β and α particles (also γ rays) frees electrons in the so called 'valence band' of the solid, with sufficient energy to jump to the 'conduction band' separated from the former by an energy gap of about 10 eV. Most of these electrons return immediately to their ground state in the valence band, but some get trapped for long times, up to several million years at ambient temperature, in what are known as crystal defects. These defects may be visualized as wells of varying depths. Natural crystals may contain up to 10^{15} to 10^{17} of such defects or traps per unit volume, produced mainly during growth. Spurious cationic impurities like Mn, Fe, Ni, and Co also represent electron traps with long lifetimes. The number of occupied traps will increase with time due to ionizing radiations from U, Th, and K, both within the crystal and from outside.

5.15.1 Thermoluminescence dating

If a crystal with occupied traps is heated, the electrons in the traps will be elevated to the conduction band from which they will fall to luminescence centres, emitting energy in the visible portion of the spectrum. In Thermoluminescence (TL) dating, a sample is heated at increasing temperatures and its light output is recorded as a glow curve, the area under which being a measure of the radiation dose received by the sample since it was last heated to ca450°C (Atken, 1985). The annual dose is determinable from the U, Th, and K concentration in the sample. Prerequisite for using the number of traps occupied by electrons (accumulated radiation dose) for dating is its proportionality to time.

Age = Accumulated radiation dose/annual radiation dose,

where annual radiation dose rate includes any external component and the intrinsic internal component. Readers are referred to original references for details of experimental techniques (Faure and Mensing, 2005; Dickin, 2005).

In Optically Stimulated Luminescence (OSL) dating, only light sensitive traps are sampled using a low power laser instead of bulk heating as in TL dating (Aitken, 1998). A third method of measuring the number of trapped electrons is to detect the absorption of microwave radiation by these electrons when immersed in a strong magnetic field using an Electron Spin Resonance (ESR) spectrometer (Grun, 1989).

In the vast majority of radioactive decays the product nuclide, as mentioned before, has too little recoil or kinetic energy to displace lattice atoms in its very short path before coming to rest. But, in the very rare type of radioactivity, namely 'spontaneous fission' (Chapter 2), the parent nuclide splits into two fragments of roughly equal mass with prodigious total energy of ~200 MeV. With so much energy the two fission fragments force their way in opposite directions through the host crystal for many lattice spacings, leaving a 'fission track' of displaced atoms, ~15 micrometre in length and ~ 100 nanometre in diameter (Figure 5.13a). The displaced atoms will remain displaced for geologically long times only in nonconducting solids. They will return to their original positions and the tracks will anneal even in a nonconducting solid when heated above a threshold temperature. ^{238}U is the only isotope in terrestrial materials to manifest spontaneous fission. An overwhelming number of ^{238}U nuclides will α decay to ^{234}Th with a half-life, $\lambda_\alpha = 1.551 \times 10^{-10}$ y^{-1}. A miniscule number will undergo spontaneous fission with a half-life, $\lambda_f = 8.46 \pm 0.06 \times 10^{-17}$ y^{-1}. Thus, only one fission will occur during the α decay of ~5 × 10^7 atoms of ^{238}U.

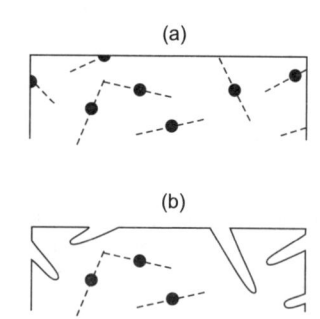

Figure 5.13 Latent fission tracks in a solid (a), and surface-intersecting fission tracks enlarged by chemical etching (b) (Geyh and Schleicher, 1990).

Highly energetic heavy charged particles also leave such trails of displaced atoms in their wake, and were examined by Silk and Barnes (1959) under a Transmission Electron Microscope with a magnification of ~ 50,000. Natural materials will contain only U fission tracks. Price and Walker (1962) showed that fission tracks that intersect a polished surface of the sample can be enlarged by etching the surface in suitable chemicals, to be seen and individually counted under an optical microscope (Figure 5.13b).

5.15.2 Fission-track (FT) dating

The number of Fission Tracks (FTs) in a natural sample can be used like radiogenic nuclides to calculate its age. From Equation (5.4a), the number of tracks accumulating in a solid over time t can be written as:

$$(sft)_v = \frac{[\lambda_f]}{[\lambda_f + \lambda_\alpha]} {}^{238}\text{U} [\exp(\lambda_f + \lambda_\alpha)t - 1] \tag{5.65}$$

where, $(sft)_v$ is the volume density of spontaneous fission tracks, and ^{238}U the present day volume concentration of this isotope in the sample. Figure 5.13b shows that only a fraction q of these tracks that intersect a polished interior surface of the sample can be etched in chemicals and enlarged for optical viewing and counting. The number of surface-intersecting tracks, $(sft)_s$, will be:

$$(sft)_s = q(sft)_v \tag{5.66}$$

Once the number of spontaneous FTs has been counted, they are annealed out by heating the sample and then it is irradiated with thermal neutrons to selectively induce fission of the companion isotope, ^{235}U. The surface density of such newly produced induced FTs will be:

$$(ift)_s = q(ift)_v = q\,^{235}U\sigma\phi \tag{5.67}$$

where, ^{235}U is the present concentration of this isotope in the sample, σ the cross section for induced fission of ^{235}U by thermal neutrons, and ϕ the dose of thermal neutrons.

Solving for t (Price and Walker, 1963) gives:

$$t = \left(\frac{1}{\lambda_\alpha}\right)\ln\left[1 + \left(\frac{\lambda_\alpha}{\lambda_f}\right)\left(\frac{sft}{ift}\right)_s \left(\frac{^{235}U}{^{238}U}\right)\phi\sigma\right] \tag{5.68}$$

ϕ is determined simply by co-irradiating glasses of known U concentration with the sample to be dated. The reader is referred to many references cited in Faure and Mensing (2005) on experimental practice, applications and limitations of FT dating.

Because fission is a very rare event in terrestrial material, FT dating is generally restricted to U-rich minerals like apatite, zircon, and sphene. Reliable ages up to 1Ga have been obtained by this method. Fission tracks, however, disappear with time, especially if the sample is heated. The age is the time since the sample was last heated above the temperature at which tracks anneal and disappear. Different minerals anneal at different temperatures, so that FT dates on several minerals from the same rock can reveal its cooling history.

The TL and FT methods have been referred to as radiation damage methods. It is more appropriate to refer to them as based on indirect radiogenic effects.

5.16 Conclusion

Initiated by unorthodox physicists and chemists in the beginning of the last century and viewed more as a curiosity until the 1950s by the Earth Science community, applications of naturally occurring parent-daughter systems now cover many modern research topics in Earth Science, such as:

1. Planetary formation: condensation and accretion of planetesimals and their early magmatic differentiation;
2. Precise chronology of: continental systems, crustal deformation episodes, pressure-temperature history of metamorphic rocks, catastrophic events;
3. Process rates for: creation of earth's crust, evolution of sea water, and genesis of hydrocarbons;
4. Understanding the processes in: mantle convection, mass transfer between crust, mantle and core, ore deposits, hydrothermal systems; and
5. High resolution chronology of very recent environmental changes.

Some of these applications will be described in later chapters.

6 Mass Spectrometry and Isotope Geochemistry

> Much of the most promising modern Earth Science depends on the measurements and observations that can no longer be obtained with a hammer, Brunton compass, and hand lens alone. Although much that is important is still being done and will continue to be done in this way by first class minds, new expensive and frequently unavailable instrumentation is called for
> **Preston Cloud.**

6.1 INTRODUCTION

Basic analytical measurements in isotope geochemistry are absolute and relative abundances of isotopes of an element available in a solid, liquid, or gaseous form. X-ray Fluorescence (XRF) Spectrophotometer, Atomic Absorption Spectrophotometer (AAS), Inductively Coupled Plasma Atomic Emission Spectrophotometer (ICP-AES), and Electron Probe Micro Analyzer (EPMA) are commonly used for elemental analysis (Potts, 1987). These instruments rely on the excitation of orbital electrons and detection of the resulting emission of radiation, or absorption of external radiation, characteristic of each element. Isotopes of an element have the same electronic structure (Chapter 1), and, hence, are not distinguished by the above instruments. Mass sensitive instruments, called mass spectrometers, have become the preferred instruments for isotope geochemistry (De Laeter, 1998). Many types of mass spectrometers have been developed over the last century, but most of them consist of three main components:

1. An ion source for production, acceleration, and collimation of ions of different isotopes of an element(s);
2. A mass analyzer for separation of ions according to their mass/charge (m/e) ratio in space or time; and
3. A detector for the collection and measurement of mass-separated ions, sequentially or simultaneously.

Mass Spectrometry and Isotope Geochemistry

6.2 Principles of Mass Spectrometry

Figure 6.1 is the schematic of a very old design of a mass spectrometer (Dempster, 1918) showing all the three basic components. Consider a beam of two singly charged isotopic ions '*a*' and '*b*' of mass m_a and m_b ($m_a + \delta m$), respectively, produced with negligible kinetic energy in front of a stack of metal plates with narrow rectangular slits, known as the ion source. The beam is accelerated by falling through a potential of *V* volts distributed between the plates of the ion source, and collimated to emerge as a narrow rectangular beam from the final slit of width, w_s, called the Source Slit (SS). The ion beam with the same kinetic energy, eV, for all ions is acted on by a uniform magnetic field, *B* gauss, directed perpendicular to the *x-y* plane (plane of the paper).

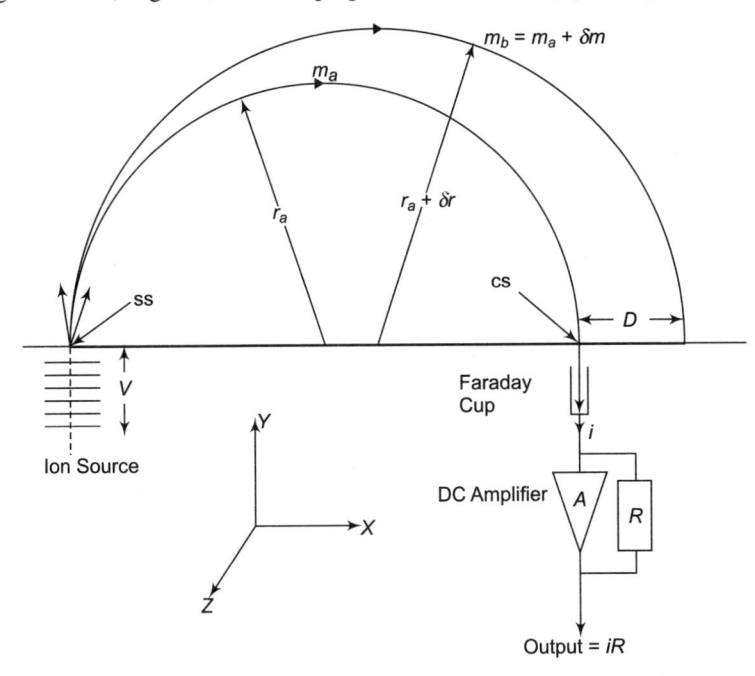

Figure 6.1 Schematic of an old design of a mass spectrometer, showing the three main components.

Kinetic energy of ion a is $eV = \left(\dfrac{1}{2}\right) m_a v_a^2$, (6.1)

where *e* is the electronic charge, and v_a the velocity of the ion. Each ion will describe a circle of radius r_a with the centripetal force due to B, Bev_a equal to the centrifugal force, mv_a^2/r_a.

$$Bev_a = \frac{m_a v_a^2}{r_a}. \tag{6.2}$$

Substituting for v_a from Equation (6.1) gives the radius as:

$$r_a = \left(\frac{1}{B}\right)\left(\frac{2Vm_a}{e}\right)^{0.5} = \left(\frac{144}{B}\right)(m_a V)^{0.5} \tag{6.3}$$

where r_a is in cm, B in gauss, and m_a in atomic mass units for singly charged ($e = 1$) ions. Example: $r_a \sim 30$ cm for a singly charged ion of mass 90 amu accelerated to 8000 V in a magnetic field of 4000 gauss. As r_a is proportional to the square root of m_a, ions with the larger mass, $m_b(m_a + \delta m)$ will describe a larger radius, r_b ($r_a + \delta r$). From Equation (6.3),

$$2 \times \left[\frac{(r_a + \delta r)}{r_a}\right]^2 = \left[\frac{(m_a + \delta m)}{m_a}\right]$$

Neglecting second order terms, $\dfrac{\delta r}{r_a} = \dfrac{\delta m}{m_a}$. (6.4)

The largest spatial separation, D, between the paths of the two ions is after a deflection of 180°. It is called the mass dispersion and is given by:

$$D = 2\,[(r_a + \delta r) - r_a] = 2\delta r = r_a \left(\frac{\delta m}{m_a}\right). \qquad (6.5)$$

D is ~ 3 mm for unit mass difference at $m = 200$. Note that the two ions' beams will converge to their starting point after one complete revolution. Consider now the ions leaving the SS at a small angle to the axial beam on either side, as indicated by the two short arrows. Ions leaving on the right of the axially directed ions will cover a little less than a semicircle and those on its other side a little more than a semi circle before converging not exactly on the axial beam, but very close to it. This approximate refocusing of initially divergent ions is called first order direction focusing in the radial plane. As this direction focusing is not perfect, the width of the focused beam is not equal to the width, w_s, of the SS, but slightly larger. Note that there is no focusing for ions diverging in the perpendicular y-z plane. If a slit with a width slightly larger than the beam width is provided at the focal point, all ions of mass m_a will pass through this slit, known as the Collector Slit (CS).

6.3 Ion Detectors

Ion currents in very early mass spectrometers were detected directly as currents by sensitive galvanometers. But now, ion currents, typically $\sim 10^{-11}$ A ($\sim 6 \times 10^7$ ions/s), are collected in a Faraday cup (a narrow mouthed rectangular metal box) located just behind the CS, and connected to a solid state direct-current amplifier with negative feedback through a high ohmic resistor, $R(10^{11}$ ohm), as shown in Figure 6.1. An ion current, i collected by the Faraday cup is converted into an exactly equal electron current in the amplifier, and appears as a proportional voltage, $v = -iR$ at the output, which can be measured with a digital voltmeter for recording and further processing. Unavoidable noise voltages in the amplifier input set $\sim 10^{-13}$ A as the smallest ion current that can be measured with a reasonably high signal-to-noise ratio using a Faraday cup + DC amplifier combination. As the statistical uncertainty in the collection of n ions is $\sqrt{n}$, the smallest ion beam must be measured long enough for the required precision. A scan of the magnetic field will sequentially bring ions of other masses to the CS to produce a mass spectrum, such as shown in Figure 6.2 for strontium with four isotopes (84, 86, 87, and 88). The peaks will show a flat top, as in the figure, if the entire refocused beam enters the CS for a small variation of the magnetic

Mass Spectrometry and Isotope Geochemistry

field. Simultaneous collection of several ions beams in independent Faraday cups, as incorporated in modern mass spectrometers, reduces analysis time and avoids errors due to drift of ion beam intensities with time.

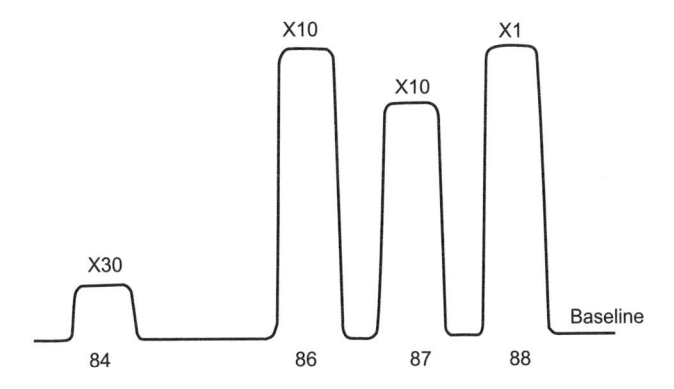

Figure 6.2 Mass spectrum of strontium. The changes in base line levels due to scale changes are not shown.

Ion currents $< 10^{-13}$ A correspond to the flow of $<10^6$ ions per second. In such cases, ions can be individually counted electronically after multiplication (by a factor of $\sim 10^4 - 10^6$) of each individual ion into a large collection of electrons in a Secondary Electron Multiplier (SEM). Such weak beams can be masked by the tail of a very intense beam one mass unit away. This tail can be reduced to less than 10 parts per billion (ppb) of the intense beam (referred as an abundance sensitivity of 10 ppb) using specially designed energy filters (see later), before detection. Using such filters, ratios as small as 10^{-6} have been measured in conventional low energy mass spectrometers. For still lower ratios, an Accelerator Mass Spectrometer (AMS) is the only choice (see later).

6.4 Sequential vs Simultaneous Detection of Ion Beams

With only one ion detector in a mass spectrometer [Figure 6.3(a)], ion beams must be measured sequentially in time, that is, one after the other. The voltage output, v for an ion beam, i is:

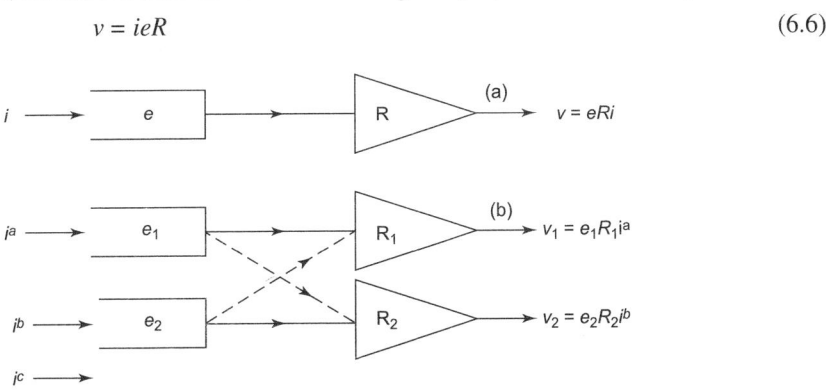

$$v = ieR \tag{6.6}$$

Figure 6.3 Single (a) and multiple collections (b) of ions.

In Figure 6.1, the efficiency, e (not to be confused with ionic charge) of conversion of ion current into an electronic current is taken as unity, but it can be slightly different for various reasons. R is the feedback resistor of the DC amplifier. The amplifier plus resistor combination serves as a transducer to convert an electron current signal into a voltage signal. Sequential outputs for two beams, i^a and i^b will be:

$$v^a = i^a e_1 R_1, \quad \text{and} \quad v^b = i^b e_1 R_1 \tag{6.7}$$

Therefore, $(i^a/i^b) = (v^a/v^b)$, as the eR factors cancel out. Sequential detection of two beams will, however, require twice as much time as for just one beam plus the time to switch, or scan the beams into the single collector. If the beam intensity drifts in time, as is invariably the case, a time-drift correction must be made. If the two beams can be simultaneously collected in two different detectors [Figure 6.3(b)], the measurement time can be reduced and drift correction eliminated.

$$v_1^a = i^a e_1 R_1, \quad \text{and} \quad v_2^b = i^b e_2 R_2 \tag{6.8}$$

$$\left(\frac{i^a}{i^b}\right) = \left(\frac{v_1^a}{v_2^b}\right)\left(\frac{e_2 R_2}{e_1 R_1}\right). \tag{6.9}$$

The collector efficiency of both Faraday cups and the gain factor of both amplifiers must be accurately determined or calibrated prior to ion current measurement. This is called static multi (dual, in this case) collection, and was first introduced in the 1940s for detecting very small differences in the isotopic composition of light elements in stable isotope geochemistry (Nier, 1947). Modern mass spectrometers incorporate many collectors which can, in addition, be moved to accommodate many elements. The residual uncertainty in gain calibration can be eliminated by two technical strategies. The first is to interchange the two beams in two sequences to get two sets of outputs, as:

$$v_1^a = i^a e_1 R_1, \quad \text{and} \quad v_2^b = i^b e_2 R_2, \tag{6.10}$$

$$v_1^b = i^b e_1 R_1, \quad \text{and} \quad v_2^a = i^a e_2 R_2. \tag{6.11}$$

$$\left(\frac{i^a}{i^b}\right) = \left[\left(\frac{v_1^a}{v_2^b}\right)\left(\frac{v_2^c}{v_1^b}\right)\right]^{0.5}. \tag{6.12}$$

Both efficiency and gain factors cancel out, but it is impractical with magnetic analyzes to interchange the order of mass separation in space. The usual practice is, therefore, to switch the magnetic field to measure i^b in the first channel, and a third beam i^c (equally spaced in mass) in the second channel in the second sequence to get two sets of outputs as:

$$v_1^a = i^a e_1 R_1, \text{ and } v_2^b = i^b e_2 R_2, \tag{6.13}$$

$$v_1^b = i^b e_1 R_1 \text{ and } v_2^c = i^c e_2 R_2 \tag{6.14}$$

$$\left(\frac{i^a}{i^b}\right) = \left(\frac{i^b}{i^c}\right)\left[\left(\frac{v_1^a}{v_2^b}\right)\left(\frac{v_2^c}{v_1^b}\right)\right]^{0.5} \tag{6.15}$$

Both gain factors and cup efficiency factors cancel out, but the measurement of the desired (i^a/i^b) ratio requires a precisely known value of (i^b/i^c). This is known as dynamic multi (dual) collection. Measurement of a third beam is not necessary if the electrical connections of the amplifiers to the cups can be interchanged (as shown by the dashed lines) between the two sets of measurements. Then:

$$v_1^a = i^a e_1 R_1 \quad \text{and} \quad v_2^b = i^b e_2 . R_2, \tag{6.16}$$

$$v_1^b = i^b e_2 R_1 \quad \text{and} \quad v_2^a = i^a e_1 R_2. \tag{6.17}$$

$$\left(\frac{i^a}{i^b}\right) = \left[\left(\frac{v_1^a}{v_2^b}\right)\left(\frac{v_1^b}{v_2^a}\right)\right]^{0.5} \left(\frac{e_2}{e_1}\right) \tag{6.18}$$

Only gain factors cancel out, but not the cup efficiency factors. However, as cup factors remain constant over long times, they do not affect measurements during relatively short analytical sessions. The trade name for amplifier interchange, or rotation, is virtual amplifier mode. Precisions of isotope ratios, as high as a few parts per million, have been realized using this scheme. Extension of these techniques to simultaneous collection of more than two beams is straight forward.

6.5 IMPROVED MASS SPECTROMETERS

Early mass spectrometers with such a large (180°) deflection angle required massive and power-intensive magnets and also location of the ion source and detector assemblies inconveniently close to the edge of the magnetic field. Nier (1940, 1947) showed that relatively low power, sector (60° or 90°) magnets also possessed direction focusing features, and provided the additional technical advantage of locating the ion source and ion detector assemblies away from the edge of the magnetic field. A non-normal angle of entrance of ion beams into 90°-sector magnetic fields was later found to provide direction focusing in both *x-y* and *y-z* planes (stigmatic focusing) for optimal transmission of ion beams, and larger mass dispersion.

If the ions from the source are not strictly monoenergetic, a magnetic analyzer will show energy dispersion in addition to mass dispersion in the focal plane, thereby degrading the mass resolution possible. A cylindrical electrostatic field analyzer also shows energy dispersion. So, a cylindrical electrostatic analyzer and a sector magnetic analyzer can be combined so that the energy dispersion of one is balanced by that of the other (Duckworth et al., 1986). If the combination is also direction focusing, the final ion image will be independent of slight variation of both, velocity and direction of ions from the source. Such double (energy and direction) focusing mass spectrometers are also essential for applications requiring high mass resolution. Quadrupole and Time-of-flight mass spectrometers (Duckworth et al., 1986) do not require magnets, but are, as yet, unsuitable for precise isotopic analysis.

The trajectories of ion beams (~2 metres) must be enclosed in a highly evacuated flight tube to avoid their collision with residual air molecules. Typical vacuum in modern mass spectrometers is less than 10^{-11} of atmospheric pressure, and achieved using turbomolecular- and ion pumps. Until the 1960s, operation, control, and data acquisition were entirely manual and laborious, which also limited the precision of isotope ratio measurements. Wasserburg et al. (1969) reported a 'home-made' mass spectrometer with automatic data acquisition and an order of magnitude higher precision. The availability of inexpensive desktop computers since the 1980s has led to full automation of commercial mass spectrometers. Multiple samples mounted on a carousel can be programmed for analysis in any sequence, night and day. The internet now allows remote monitoring and control of the machine. Indispensable though they have become, these sophisticated data systems render the basic mass spectrometers somewhat opaque to most users.

6.6 Types of Ion Sources Used in Isotope Geochemistry

6.6.1 Thermal ionization source

A fraction of atoms of an element evapourating from a hot metal surface is positively ionized according to the Saha-Langmuir equation:

$$\left(\frac{n^+}{n^o}\right) \sim \exp\left[\frac{e(W-IP)}{kT}\right], \quad (6.19)$$

where e is the electronic charge; W work function of the metal; IP ionization potential of the element; k Boltzmann constant, and T the absolute temperature. Tantalum, rhenium, and tungsten are metals with a high operating temperature and large W. Elements with $IP < W$ will ionize efficiently. A schematic diagram of a thermal ionization source is shown in Figure 6.4(a). Ions, emitted from the filament surface with negligible energy spread, are collimated into a rectangular ion beam for mass analysis. One of the limitations of the thermal ionization source is the lack of sensitivity for elements with high IP. Of the many empirical recipes for efficient ionization of such elements, a silica gel plus phosphoric acid mixture on a rhenium filament has led to dramatic increases for lead (Cameron et al., 1969). Another technique is to use two adjacent filaments: one at a low temperature for evapouration, and the other at a higher temperature for efficient ionization (Inghram and Chupka, 1953).

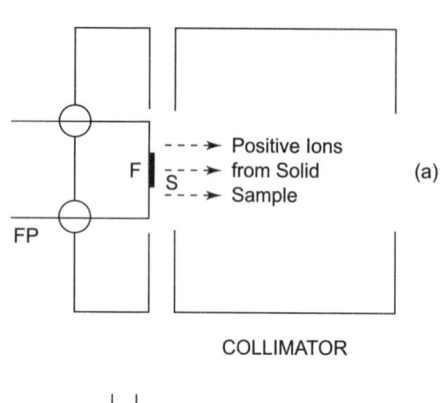

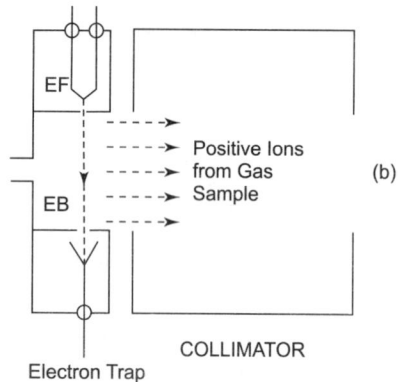

Figure 6.4 Schematic diagram of a thermal ionization source for solid samples (a), and that of an electron impact ionization source for gaseous samples (b).

Elements with high Electron Affinity (EA) ionize better as negative ions from a hot metal surface with low W according to the modified Saha-Langmuir equation (Heumann et al., 1995):

$$\left(\frac{n^-}{n^o}\right) \sim \exp\left[\frac{e(EA-W)}{kT}\right] \quad (6.20)$$

Lanthanum and barium salts on a metal surface reduce its effective W. Efficient ionization of Re and Os as negative metal oxides facilitated the application of Re-Os dating (Creaser et al., 1991).

6.6.2 Electron impact ionization source

Developed mainly to ionize gaseous elements and molecules, this source uses a ~70 eV electron beam [EB in Figure 6.4(b)] to ionize gaseous atoms or molecules in its path. The ions produced are collimated into a rectangular ion beam with a negligible energy spread as in a thermal ionization source. As the ion beam will include ions from the residual atmospheric gases close to the ion source, mass spectrometers must be pumped to ultra high vacuum to reduce such ions to negligible levels.

6.6.3 Secondary ionization source

In this ion source, a finely focused primary high energy primary beam of cesium or oxygen ions is used to bombard a microscopic volume on a polished surface of a solid sample [S in Figure 6.5(a)] to sputter a secondary beam of ions from the volume for subsequent mass analysis [Figure 6.5(a)]; hence, the name Secondary Ionization. The sputtered beam shows considerable energy spread and includes nominally isobaric unwanted ions from the sample.

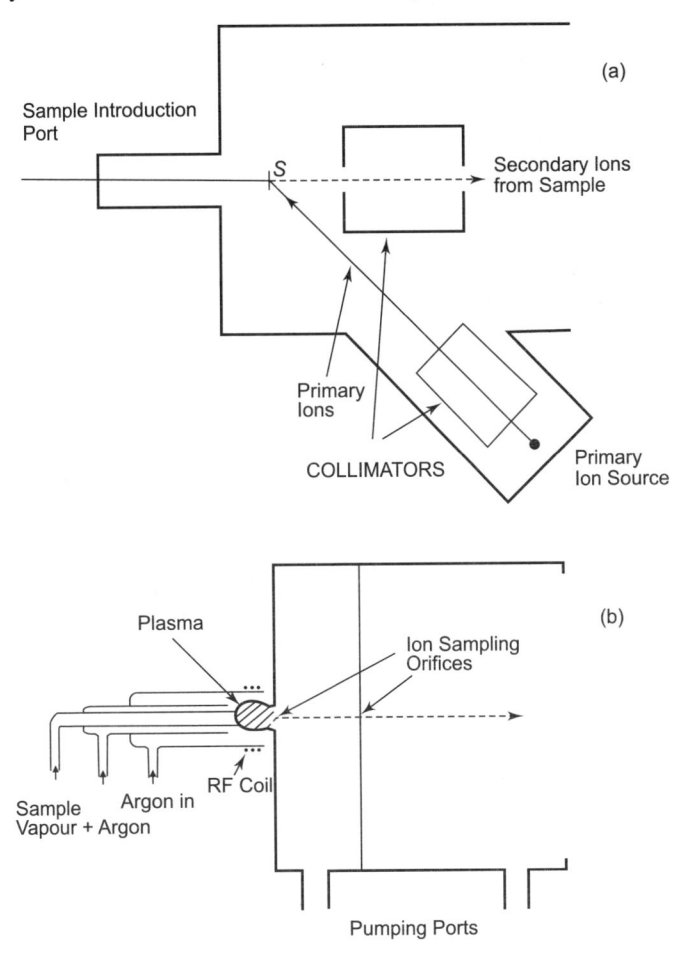

Figure 6.5 Schematic diagram of a secondary ion source (a), and that of an inductively-coupled plasma ion source (b)

6.6.4 Plasma ionization source

Radio Frequency (RF) electrical energy from a RF generator, coupled to a flowing argon gas stream, creates a narrow zone of very high temperature plasma. Elements nebulized into this zone (~10,000 K) are efficiently ionized [Figure 6.5 (b)]. Elements can be introduced into the plasma, either as an aqueous vapour from a solution, or as a gas from vapourizing solid samples usually with a laser beam. However, ions produced in this (ICP) ion source show considerable spread in energy, and fluctuations in intensity.

6.7 Typical Commercial Mass Spectrometers Using Different Ion Sources

A schematic of the complete ion optical system of a modern commercial mass spectrometer with a thermal ionization source (TIMS, for short) is shown in Figure 6.6. As ions from this source are homogeneous in energy, but vary slightly only in direction, a single 90° sector magnet is adequate for both mass separation and direction focusing, as discussed in Section 6.2. Non-normal incidence of ion beam on the magnet facilitates direction focusing both in and perpendicular to the plane of the paper (stigmatic focusing). The mass dispersion of the ~ 25 cm radius magnet is adequate to locate four Faraday cups to collect paraxial ion beams, as shown. The cups are moveable precisely along an oblique focal plane. The axial beam is measured a little beyond its collector slit as a current by being electrically deflected into the axial Faraday cup offset to one side or counted as individual ions in a SEM plus counter. If the axial beam is several orders of magnitude (typically 6 or 7) smaller than an adjacent isotope beam, the tailing effect of the latter on the former must be reduced by at least this factor. This is achieved by passing it straight through an energy (called Retarding Potential Quadrupole) filter, which cuts off ions with the wrong energy. Such filters can realize abundance sensitivities of 10 ppb.

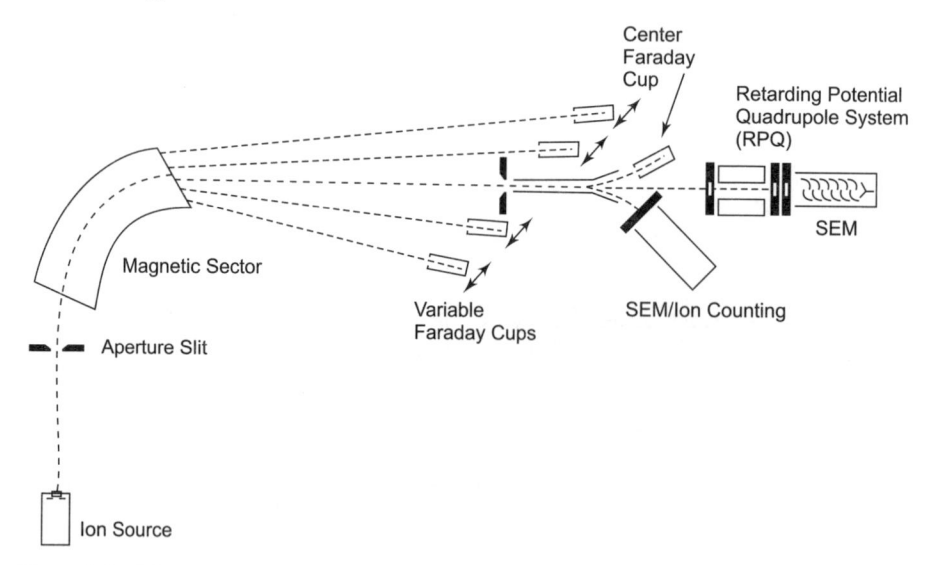

Figure 6.6 Schematic of a commercial version (Finnigan MAT 262 RPQ) of a Thermal Ionization Mass Spectrometer with extended geometry ion optics and five collectors. The centre beam can be deflected into either a Faraday cup, SEM, or a RPQ filter.

The design of a commercial mass spectrometer with a secondary ion source (SIMS, for short) is shown in Figure 6.7 (De Laeter, 1990). The secondary ion source in the lower right can be easily recognized from Figure 6.5(a). As the secondary ion beam is inhomogeneous both in energy and direction, the mass analyzer should be capable of correcting both. The combination of a electrostatic analyzer and sector magnet provides such double focusing capability. The multi-collector array is for simultaneous collection of mass-separated secondary ions. The second electrostatic analyzer serves the same purpose as the Retarding Potential Quadrupole (RPQ) in Figure 6.4 to measure an extremely scarce axial beam in the presence of an abundant isotope. The secondary beam will consist of atomic ions of the desired element and nominally isobaric molecular ions of other elements in the minute sample volume penetrated by the primary ion beam. Therefore, the mass analyzer must also have a sufficiently high mass resolution to discriminate the wanted ions from the unwanted.

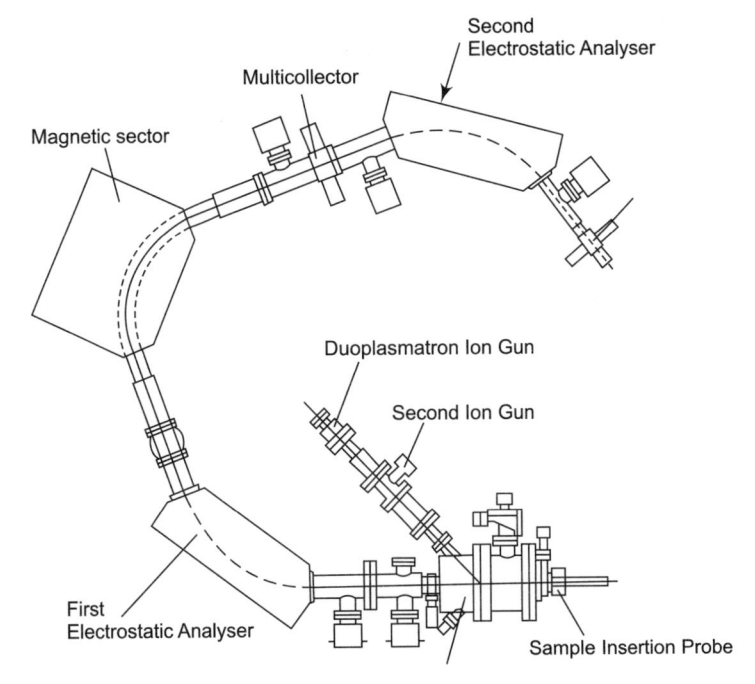

Figure 6.7 Schematic of a SIMS at Cambridge University.

W. Compston and his colleagues at the Australian National University realized even in the early 1970s that a SIMS-type mass spectrometer with a high sensitivity and mass resolution would be ideal for isotopic analysis of minerals with high spatial resolution (a few tens of microns) and without the need of prior chemical processing. They, therefore, designed and built a Sensitive High Resolution Ion Micro Probe (SHRIMP) using a very intense primary ion beam (from a duoplasmatron source) and large geometry ion optics (Compston et al., 1984). This has been very successful in U-Pb dating of zircons and other uranium-rich minerals.

In the conventional design of a SIMS, the primary and secondary beams are not coaxial (Figure 6.7) which, for practical reasons, imposes a compromise between the smallest primary beam and collection of secondary ions from sample. In a modern variant of the conventional SIMS (called NanoSIMS), the two beams are designed to be coaxial close to the sample, resulting in an extremely fine primary beam for micron scale sampling, and high collection efficiency of sputtered or secondary ions. This instrument has been very critical for the isotopic analysis of micron-size presolar grains in primitive meteorites (Chapter 8).

The incorporation of an inductively-coupled plasma ion source with its high ionization efficiency for most elements in a mass spectrometer posed a new technical problem not met with in other ion sources, namely ions produced in argon plasma at atmospheric pressure and high temperature must first be extracted into the high vacuum of a mass spectrometer. The front end of a successful commercial version of a mass spectrometer (Figure 6.8) shows how this problem is tackled. Ions from the high temperature plasma at atmospheric pressure are first extracted through a fine hole (1 mm diametre) in the first cone (extraction cone) immersed in the plasma into a region separated from the rest of the mass spectrometer by another on-line cone with a still smaller orifice. The bulk of neutral argon gas rushing into this region is pumped out by a high capacity fore pump to a steady vacuum of 10^{-4} mm of mercury. The ions streaming from the first cone to the second cone enter the very high vacuum region of the mass spectrometer evacuated by turbomolecular pumps. The electrostatic analyzer and sector magnet is a standard double focusing combination (as in Figure 6.7) to focus ions of the same mass, but differing in energy and direction. Finally, simultaneous collection of mass-separated ions in a multi-collector array solves the fluctuations in beam intensity due to the inherent plasma instability.

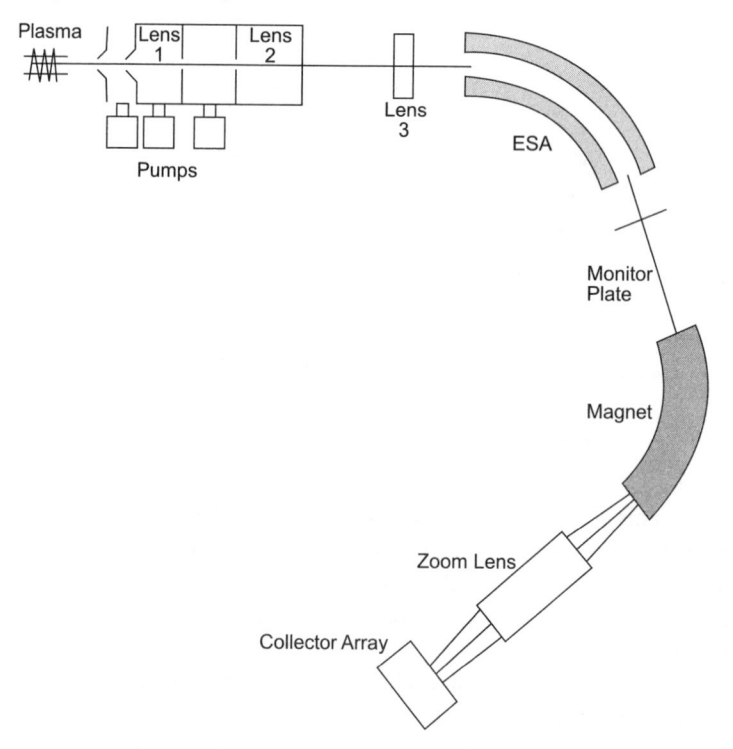

Figure 6.8 Schematic of a MC-ICP-MS made by Nu Instruments.

In contrast to the conventional moveable collectors, this design uses a fixed array of collectors and a zoom lens just in front of the array to align the separated ions into their respective Faraday buckets in the array. This greatly simplifies the mechanical and electronic complexity of moveable collector arrays. Multi-collection is a must to integrate unstable ion currents long enough for high precision isotopic ratio measurements comparable with TIMS. For less precise measurements, a single ion-counting detector would suffice, if the required mass range can be scanned at a much higher frequency than that of plasma fluctuations. Very early ICP-MS instruments indeed used a fast scanning quadrupole mass analyzer (no magnet) followed by an ion counter (Potts, 1987). The two lens stacks in Figure 6.8 also serve to match the circular image of the extracted beam into the rectangular defining slit of the mass analyzer combination. Details of many commercial multi-collector inductively coupled plasma mass spectrometers (MC-ICP-MS) can be found in Halliday et al., 2000; Rehkamper et al., 2000.

6.8 Mass Fractionation in Mass Spectrometers

Mass fractionation, mass discrimination, or machine bias refers to the deviation of the ratios of instantaneous ion currents from the true ratios of the isotopes in the sample at that time. It is due to preferential evapouration of lighter isotopes. While this fractionation is primarily mass dependent, its exact dependence on mass is very complicated (Habfast, 1983; Hart and Zindler, 1989). Three empirical relationships have been developed to correct for fractionation to achieve increased reproducibility in isotopic analysis of elements relevant in isotope geochemistry. These are (Wasserburg et al., 1981):

$$\left(\frac{a}{b}\right)_c = \left(\frac{a}{b}\right)_o [1 + \alpha(m_a - m_b)] \quad \text{Linear law.} \quad (6.21)$$

$$\left(\frac{a}{b}\right)_c = \left(\frac{a}{b}\right)_o [1 + \beta]^{(m_a - m_b)} \quad \text{Power law.} \quad (6.22)$$

$$\left(\frac{a}{b}\right)_c = \left(\frac{a}{b}\right)_o \left[\frac{m_a}{m_b}\right]^{\gamma m_b} \quad \text{Exponential law.} \quad (6.23)$$

where $(a/b)_o$ is the observed fractionated ratio of two isotopes of mass m_a and m_b, respectively, $(a/b)_c$ is the corrected (not necessarily the true) ratio in the sample, and α, β, γ are the fractionation coefficient per unit mass in the three cases. Exponential law seems closest to actual process. As fractionation is generally variable in time and between samples, it cannot be predetermined.

If at least one isotope ratio in an element is constant and known, it can be used as an internal standard (see Section 5.3) to correct its other isotope ratios for machine bias according to one of the above laws. For strontium, a constant ratio of its two non-radiogenic isotopes, $^{88}Sr/^{86}Sr$ assumed to be exactly 8.375321, is used to correct the ratio of interest, $^{87}Sr/^{86}Sr$, to realize a very high precision (~5 ppm). ($^{88}Sr/^{86}Sr$) is called the normalizing ratio, and ($^{87}Sr/^{86}Sr$) is called the normalized ratio. In the absence of any such internal standard, a known combination of two artificially enriched isotopes (called double spike) is mixed with the sample element to monitor machine bias. As three of the four stable isotopes of lead are radiogenic and, hence, variable, a ^{204}Pb-^{207}Pb double spike is used for high precision lead isotopic analysis (Compston and Oversby, 1969). Machine fractionation is unavoidable in any mass spectrometer developed so far. It must be measured in each case to realize very high precision or external reproducibility

6.9 ABSOLUTE ABUNDANCE OF AN ISOTOPE

Elemental analysis is invariably accomplished with reference to a calibration curve defined by a set of standard solutions of the element(s) of interest. While this approach is possible with some mass spectrometers to determine the absolute abundance of an isotope of an element, the preferred method is known as Isotope Dilution Mass Spectrometry (IDMS).

6.9.1 Isotope dilution mass spectrometry

This method depends on the measurement of the change in the isotopic ratio of an element by the addition of a known quantity of the same element, but with a different isotopic ratio (artificially produced) called the spike or tracer (Heumann, 1992). Such enriched isotopes are available commercially and some could be very expensive.

If the mole quantity of an isotope a is $(a)_s$ in the sample, $(a)_t$ in the tracer and $(a)_m$ in the mixture, then:

$$(a)_m = (a)_s + (a)_t \tag{6.24}$$

Similarly, for another isotope b,

$$(b)_m = (b)_s + (b)_t \tag{6.25}$$

$$(b)_m \left(\frac{a}{b}\right)_m = (b)_s \left(\frac{a}{b}\right)_s + (b)_t \left(\frac{a}{b}\right)_t \tag{6.26}$$

Substitute for $(b)_m$ to get:

$$(b)_s = (b)_t \left[\frac{\left(\frac{a}{b}\right)_m - \left(\frac{a}{b}\right)_t}{\left(\frac{a}{b}\right)_s - \left(\frac{a}{b}\right)_m} \right] \tag{6.27}$$

All items in the right side are either known or measurable. Thus, the total mole quantity of the normal element with only two isotopes is:

$$(a)_s + (b)_s = (b)_s \left[1 + \left(\frac{a}{b}\right)_s \right] \tag{6.28}$$

The most critical requirement for IDMS is to ensure complete homogenization of the sample and tracer isotopes in the mixture. A great advantage of IDMS is that after isotopic equilibration, chemical separation of the element from matrix elements need not be quantitative. The propagation of the random error in $(a/b)_m$ into $(b)_s$ is the least when:

$$\left(\frac{a}{b}\right)_m = \left[\left(\frac{a}{b}\right)_s \left(\frac{a}{b}\right)_t \right]^{0.5} \tag{6.29}$$

The IDMS is applicable to elements with at least two isotopes. Monoisotopic elements can also be measured if they have a radioactive isotope of long enough half-life.

6.10 SAMPLE SIZE REQUIREMENTS

Basic statistical precision is $\sqrt{n}$, n being the number of atoms of each isotope detected. High precision in ratio measurement translates into high resolution in time measurement. For a high precision of 0.001% (10 ppm) in the measurement of ^{87}Sr isotopes (say) in a sample, minimum sample requirement can be calculated as follows (Wasserburg, 1987):

- ~10^{10} atoms of ^{87}Sr - for 0.001% precision
- ~10^{11} atoms of elemental Sr - for ~10% ^{87}Sr in total Sr
- ~10^{13} atoms of elemental Sr- for 1% ionization efficiency
- ~10^{14} atoms of elemental Sr - for 10% counting time for ^{87}Sr.

10^{14} atoms correspond to 15×10^{-9} g of Sr or a total sample weight of 15×10^{-5} g with a typical Sr content of 100 ppm. Thus, sample requirement is extremely small even for this level of precision. With a higher ionization efficiency of 0.1 and simultaneous collection of all four Sr isotopes, minimum sample weight can be as small as one microgram: the mass of a dust particle of about 50 microns in size. However, a much higher practical limit is set by inevitable contamination during sample dissolution from lab air, water and reagents, lab ware and handling, known as blank. A mixture of sample Sr and blank Sr will be (according to Equation 6.26):

$$(^{86}Sr)_m \left(\frac{^{87}Sr}{^{86}Sr}\right)_m = (^{86}Sr)_s \left(\frac{^{87}Sr}{^{86}Sr}\right)_s + (^{86}Sr)_b \left(\frac{^{87}Sr}{^{86}Sr}\right)_b \qquad (6.30)$$

where $(^{86}Sr)_m$, $(^{86}Sr)_s$, and $(^{86}Sr)_b$ are the molar concentrations of ^{86}Sr in the mixture, sample and blank, respectively. If the blank ratio differs significantly from the sample ratio, it is clear that the blank contribution $(^{86}Sr)_b$ must be extremely small relative to $(^{86}Sr)_s$, if the mixture ratio is to be indistinguishable within its precision limits from the intrinsic sample ratio. Hence, the need for metal-free labs with finely filtered airflow, ultrapure reagents and water, teflon/plastic labware, and ultraclean chemical separation procedures and mass spectrometry (Wasserburg, 1987).

6.11 MASS SPECTROMETRY VS DECAY COUNTING

For quantitative detection of stable isotopes, the mass spectrometer is the only option. But for radioactive isotopes, there is another option, namely, counting the fraction of atoms decaying in a very large collection, p.

The number of disintegrations in a counting interval, $t = t\lambda p = (tp)(0.693/T)$ \qquad (6.31)

where p is the total number of atoms in a sample and T the half life of the species. The statistics of sampling requires that at least 10^4 atoms are detected for a precision of 1 per cent. If the fraction of the sample atoms that is ionized and detected in a mass spectrometer is 10^{-3} (a conservative value), the required number of sample atoms is 10^7 and the analysis time is a few minutes. If 10^4 atoms are to decay out of 10^7 atoms in a counting time of 24 hrs (3×10^{-3} y), the half-life of the species should be about 2 y. In other words, mass spectrometric detection is far more sensitive and much quicker than decay counting for radioactive species with half lives longer than 2 years. For this reason, mass spectrometry has replaced decay counting as a measurement technique (Chen et al., 1986; Edwards et al., 1987) in many applications.

6.12 ACCELERATOR MASS SPECTROMETER

Some of the radioactive isotopes of interest in isotope geochemistry are present in extremely small concentrations, as for example, cosmic-ray produced (cosmogenic) ^{14}C in modern carbon. One gram of carbon extracted from a living organism will contain 5.0×10^{22} stable atoms of ^{12}C, 5.6×10^{20} stable atoms of ^{13}C, and only 5.8×10^{10} atoms of radioactive ^{14}C giving an extremely small value of 1.2×10^{-12} for the desired $^{14}C/^{12}C$ ratio in the sample (Section 5.4). Isotope ratios smaller than 10^{-6} cannot, in practice, be measured with conventional low voltage mass spectrometers because of technical constraints, such as unavoidable isobaric atomic and molecular impurities in the sample, maximum usable ion current, and detector background noise. In the case of carbon, isobaric (nominal mass 14) atomic impurity is ^{14}N, and isobaric molecules are $^{12}C^{1}H_2$ and $^{13}C^{1}H$. For this reason, concentration of ^{14}C and such relatively short lived nuclides were measured until 1980 by only counting their radioactive disintegrations, usually over very long time intervals of days and weeks (Allegre, 2008).

The Accelerator Mass Spectrometer (AMS), developed in the late 1970s (Elmore and Phillips, 1984) has largely overcome the limitations of conventional mass spectrometers. Basically, it includes an accelerator with a terminal voltage of 3 million volts between two sector (90°) magnetic mass spectrometers, as shown schematically in Figure 6.9. Its principle of operation can be understood from its original and still main application to ^{14}C analysis (Figure 6.9).

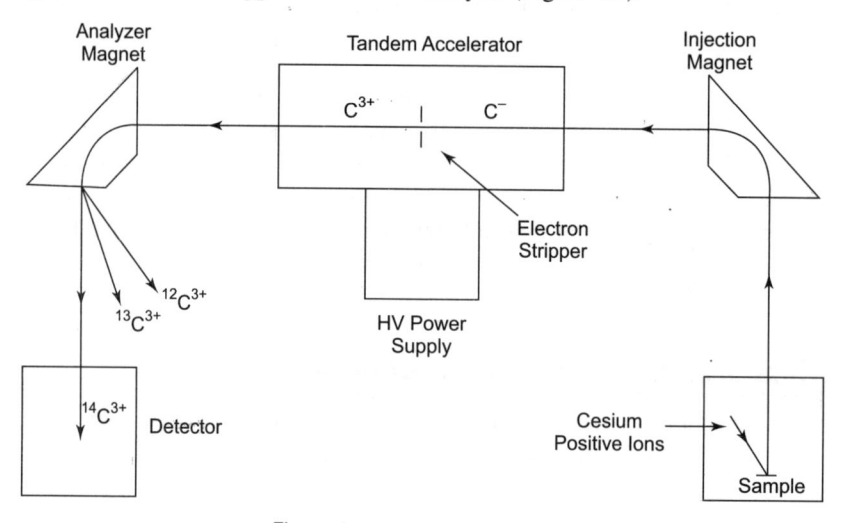

Figure 6.9 Schematic of an AMS

In the ion source, a Cs^+ ion beam sputters negatively charged carbon ions from a small button of highly purified graphite prepared from the sample of interest. As ^{14}N does not form negative ions, it is neatly eliminated from the carbon ion beam. After acceleration to a few thousand volts in the ion source, the carbon beam passes through the first magnetic sector mass spectrometer, which selects negative ions of mass 14 (^{14}C and any other interfering molecular ions of the same mass), and injects them into the tandem accelerator. With a gain in energy of 3 MeV on arrival at the central terminal, the beam passes through a thin gas cell or metal foil, which strips up to four electrons from the negative ions, thereby converting them to a positive charge with a majority in the charge state 3+.

This charge change in the gas cell is the key to the elimination of molecular ions, as monoatomic ^{14}C ions will be stable with a 3$^+$ charge, but triply charged molecules will dissociate instantaneously into smaller fragments due to excessive Coulomb repulsion. The ^{14}C^{+3} ions and fragments will now be accelerated by the positive terminal voltage down the second half of the accelerator tube to emerge with a total energy of 12 MeV to enter the second sector magnet. This magnetic analyzer tuned to mass 14 will filter out any residual fragments and pass only the monoatomic atoms of mass 14 to the detector. If ^{13}C and ^{12}C ions are sequentially injected into the accelerator after ^{14}C, they will be deflected by the second magnet into suitably positioned Faraday cups for measurement of either ^{14}C/^{13}C or ^{14}C/^{12}C ratio. The overall operation of the AMS on other rare radionuclides is the same, but with appropriate strategies to suit each (Finkel and Suter, 1993).

7 Error Analysis

> The uncertainty of a date is as important as the date itself.
>
> K Ludwig

7.1 INTRODUCTION

The calculation of a radiometric age (strictly speaking, radioactive decay interval) of a natural object is ultimately based on the decrease in the concentration of a parent isotope in that object during the interval. The measurement of the initial and final concentrations (directly or otherwise) in terms of an internationally accepted unit (moles per unit weight of the sample) is always subject to analytical errors or uncertainties. These analytical errors will be reflected (propagated) in the final age result. In this chapter we will consider generally accepted methods of estimating analytical error in a measured quantity, and how it will affect the final result based on this measurement (Cunningham, 1981).

The absolute error in a measurement is the discrepancy between the measured value and its 'true' value. Except in the case of counting discrete objects, a measurement of a physical quantity in terms of an internationally-accepted unit will always be subject to an uncertainty. The true value that we are seeking is unknowable, unless it is already known in some other way. So, it is necessary to state from the results within what range of values the true value is likely to occur and the confidence that can be placed on the validity of this statement. The statement of a result without any indication, with a certain confidence, of the range of values bracketing the true value has virtually no significance to end users. The idea of errors is, therefore, not something of only secondary or peripheral interest in an experiment. On the contrary, it is related to the purpose of the experiment, the method of doing it and the relevance of the results to a significant scientific question. The quotation in the beginning of this chapter reflects the importance of error limits to an experimentally measured date or age.

7.2 Systematic and Random Errors

Before proceeding further on error estimates, we must distinguish between two types of experimental errors—determinate or systematic, and indeterminate or random. Determinate errors generally shift the measured result in one direction with respect to the true value. They are often reproducible, and, in many cases, may be measured, and or corrected for by careful and meticulous experimentation. Examples of such errors are incorrect calibration of the measuring instrument used to measure the quantity in question, alteration of the quantity by extraneous causes like contamination, non-reproducible conditions during replicate measurements of the same quantity, and observer bias. So, determinate errors are distinguished as instrumental, methodic, and operative in accordance with their origin. Care must be taken in designing the experiment to reduce the size of the systematic error as much as possible, and then try to discover the size of the remaining significant systematic error. In what follows, the measurements are considered to be free from any systematic error.

The other error, indeterminate or random, as the name implies, cannot be attributed to any known cause. They are invariably present resulting in both high and low results with equal probability. If a quantity is measured with an instrument with good resolution (sometimes called resolution), it is found that repeated readings are not identical, as would be the case with only systematic errors. They cannot be eliminated or corrected, and are the ultimate limitation on the measurement. Fortunately, repeated readings not only reveal random errors, but also facilitate reduction of their effects in the final measured result. In contrast, repeated measurements with the same apparatus and observer neither reveal nor reduce systematic errors. In the context of errors it is important to distinguish between the terms 'accuracy' and 'precision'. A result is said to be accurate if it is relatively free from systematic error, and precise, if the random error is small.

7.3 Measurement of Random Data

Results that scatter in a random fashion are best treated by techniques of statistics—the art of determining the probable from the possible. Detailed treatment of the statistical error theory is beyond the scope of this book. We will consider only how these techniques are applied and what information they furnish beyond what may be apparent in the raw data. We will first consider how best to convey the result of random data. Listing the results in order from the lowest to the highest is more informative than listing them sequentially in order of measurement, as the former readily shows the minimum and maximum values and the middle or median value.

A graphical presentation of the results is even more informative. A particularly convenient form of graphical presentation is the histogram, such as shown in Figure 7.1. The area of each rectangular cell (seven in the figure) is proportional to the number of results within the limits given by the edges of that cell. If the widths of the cells are equal as in the figure, the height of the cell is proportional to the frequency of the results within that cell. The histogram has a jagged appearance because it represents only a few repeated measurements, as is practical in an experiment. If the number of measurements is hypothetically large, the histogram will become a smooth curve, provided the measuring instrument gives sufficiently fine readings. This is known as the distribution function. We shall not specify the exact form of such a curve at this stage, but we expect such a curve to be symmetrical about its centre, as the dotted curve superimposed on the histogram in Figure 7.1.

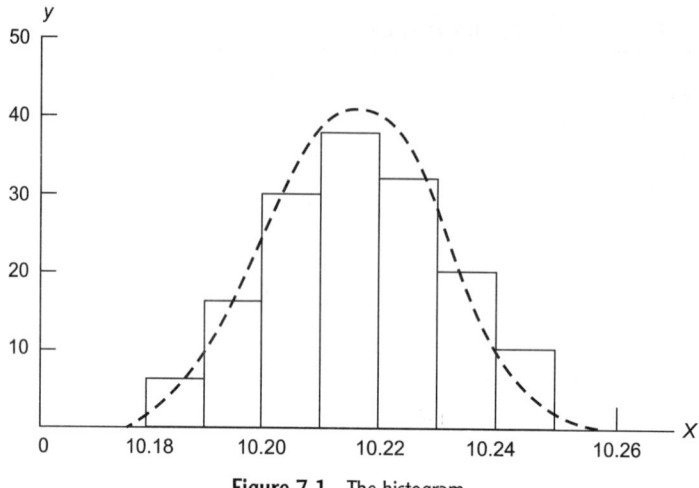

Figure 7.1 The histogram.

Although a histogram or distribution function displays much of the information in the results, it is often useful to describe the series of results by numbers for easy communication to other workers. Two numbers have become very useful in this respect—one giving the central tendency of the results and the other giving a measure of the scatter of the results. The first number is the simple arithmetic mean of the results. The mean of measurements for a given quantity carried out by the same apparatus, procedure and conditions, and observer approaches a definite value as the number of measurements increases indefinitely. In the theory of errors (Bevington, 1969) this mean is called the population or limiting mean. It is assumed to be the true value of the quantity. The other number is called the standard deviation of the measurements, and is defined as the value approached by the square root of the average of the sum of the squares of the deviation of individual results from the limiting mean, as the number of measurements increases indefinitely. In symbols:

$$\left[\frac{1}{n}\Sigma(x_i - \mu)^2\right]^{\frac{1}{2}} \to \sigma, \text{ as } n \text{ tends to infinity.} \tag{7.1}$$

σ is the standard deviation, μ the population mean, and x_i is the ith measurement. Variance is defined as the square of the standard deviation. The reason for the common use of the term 'standard' rather than the explicit term 'root mean square' is not clear.

7.4 POPULATION MEAN AND SAMPLE MEAN

In a statistical context, a sample refers to a group of few individual measurements from a much larger population of measurements and is assumed to be representative of the population. As only a few replicate measurements of the same quantity are practical, it might appear that the concepts of limiting or population mean and standard deviation are of little practical use. Fortunately this is not the case. An important theorem in theory of errors, called the Central Limit Theorem states: 'If a population has a finite variance σ^2 and mean μ, then the distribution of the means of small samples drawn randomly from the parent population approaches the normal distribution with variance σ^2/n

Error Analysis

and mean μ, as the sample size n increases. The theorem does not specify the form of the unknown population, which is good as it cannot anyway be known from a few actual measurements. The theorem does state that the distribution of sample means of independent measurements will be approximately normal about the population or limiting mean with a standard deviation for this distribution as $\sigma/\sqrt{n}$. $\sigma/\sqrt{n}$ is called the standard error of the mean. In other words, for a set of measurements with more or less symmetrical histogram and a small deviation from the mean, the distribution of sample means is very close to the normal even for n as small as 3 or 4 (Ku, 1969). This condition is satisfied by a large number of physical measurements.

The normal distribution referred to by the Central Limit Theorem is well known in statistics, and is illustrated in Figure 7.2.

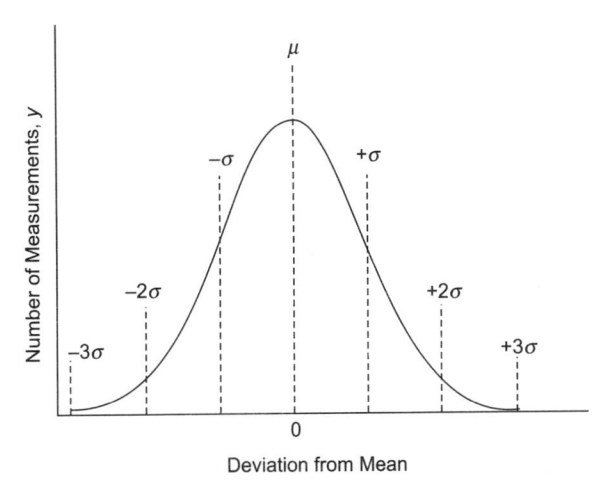

Figure 7.2 The normal distribution.

Mathematically, the normal distribution is given by:

$$y = \left(\frac{1}{\sigma\sqrt{2\pi}}\right) \exp\left[\frac{(x_1 - \mu)^2}{2\sigma^2}\right] \tag{7.2}$$

The normal distribution is one in which values cluster around a central point μ and the frequency of measurements will fall off exponentially on both sides at a rate inversely proportional to the magnitude of the other parameter σ. The relative frequency of data between x and $x + \delta x$ is $y\delta x$. y is, therefore, called the probability density of the distribution. The area under the curve between two x values is, thus, the probability that the quantity will fall between these two values. The total area under the curve is unity. Table 7.1 gives the probability that a normally distributed quantity will be within r standard deviations of the mean.

Table 7.1 Probability of Occurrence of a Normally Distributed Quantity

r	% Probability
0.6745	50
1	68.27
2	95.43
3	99.73
∞	100

It is worth repeating that the Central Limit Theorem says that to a good approximation, the mean of a fairly small sample will lie on a normal distribution with a mean equal to the population mean and standard deviation $\sigma/\sqrt{n}$, where σ is the population standard deviation. This theorem is powerful because it enables statements to be made on the probability that the population mean is less than a specified deviation from the sample mean. So, if the sample mean is $\bar{x}$ and the population mean that we are seeking is μ, then in more than 95 per cent of samples, the population mean will be within two standard deviations of the sample distribution, as given by the relation:

$$-\frac{2\sigma}{\sqrt{n}} < \bar{x} - \mu < +\frac{2\sigma}{\sqrt{n}}$$

Or,

$$\bar{x} - \frac{2\sigma}{\sqrt{n}} < \mu < \bar{x} + \frac{2\sigma}{\sqrt{n}} \tag{7.3}$$

will be correct for over 95 per cent of a large number of sets of samples. This allows statements to be made, from a single small sample of measurements, on the probability of the population mean being more than $2\sigma/\sqrt{n}$ away from the sample mean. But the snag is that it is necessary to know σ, which is the standard deviation of the population and can only be found from a very large number of measurements. Fortunately, the population standard deviation can be estimated from the sample standard deviation s, where s is given by:

$$s = \left[\frac{1}{(n-1)} \sum (x_i - \bar{x})^2\right]^{0.5} \tag{7.4}$$

Since $\bar{x}$ is found from the same n results, the last value is fixed by $\bar{x}$ s and all other $(n-1)$ results. The degree of freedom is therefore $(n-1)$. To allow for the fact that s may still be an underestimate of σ, the above so-called confidence interval has to be slightly widened as:

$$\bar{x} - \frac{ts}{\sqrt{n}} < \mu < \bar{x} + \frac{ts}{\sqrt{n}} \tag{7.5}$$

The multiplier, t, depends on n and the probability level chosen, and is called Student's t. The distribution of t for low values of n was obtained by Gosset (1908) and published under the pen name 'Student'. Values for t have been evaluated for the normal distribution and given in standard statistics books. To find t, it is necessary to know the number of degrees freedom of the calculated sample standard deviation, s.

For example, if a series of five measurements of voltage give a mean $\bar{x}$ of 5.314 v and a standard deviation s of 0.010 v, then for a 95 per cent probability level and degree of freedom 4, $t = +4.6$. The statement can, therefore, be made that there is a 95 per cent probability that the population mean μ will be in the range:

$$5.314 - \frac{4.6 \times 0.010}{\sqrt{5}} < \mu < 5.314 + \frac{4.6 \times 0.010}{\sqrt{5}} \tag{7.6}$$

Or,

$$5.293 < \mu < 5.335.$$

Error Analysis

The ability to assign a probability that the population mean will be within a certain specified range of the mean of a fairly small sample is the basis of the application of statistics to measurement. The above coverage of this very complicated subject is necessarily a simplified one. Standard text books on the subject give a detailed treatment with proofs.

7.5 PROPAGATION OF MEASUREMENT UNCERTAINTIES

Let us now consider the determination of the mean value and standard deviation of a final result u which is function of one, two, or more variables $(x, y, z, ...)$ which are actually measured with mean values $\bar{x}, \bar{y}, \bar{z}, ...$ and standard deviations $\sigma_x, \sigma_y, \sigma_z, ...$, respectively. For simplicity of calculations, let u be a function of only two variables, x and y. That is:

$$u = f(x, y) \tag{7.7}$$

The most probable or mean value of u can be calculated as

$$\bar{u} = f(\bar{x}, \bar{y}) \tag{7.8}$$

The uncertainty in u can be found by considering the spread of values resulting from combining the individual measurements x_i and y_i into individual results for u:

$$u_i = f(x_i, y_i) \tag{7.9}$$

Using elementary calculus (Bevington, 1969) the deviations $(u_i - \bar{u})$ can be expressed in terms of the deviations $(x_i - \bar{x})$ and $(y_i - \bar{y})$ as:

$$(u_i - \bar{u}) = (x_i - \bar{x})\left(\frac{\delta u}{\delta x}\right) + (y_i - \bar{y})\left(\frac{\delta u}{\delta y}\right) \tag{7.10}$$

The terms $(\delta u/\delta x)$ and $(\delta u/\delta y)$ are called partial derivatives of the function u with respect to x and y, respectively. They are defined as proportionality constants between changes in u and changes in x and y evaluated at the point $(\bar{x}, \bar{y})$ for infinitesimal changes in x and y. From the definition of standard deviation given in Equation (7.2) we can express the standard deviation σ_u as:

$$\sigma_u^2 = \lim\left(\frac{1}{n}\right)\Sigma(u_i - \bar{u})^2 \text{ in the limit } n \text{ tending to infinity.}$$

From Equation (7.10) we can express the variance σ_u^2 as:

$$\sigma_u^2 = \lim\left(\frac{1}{n}\right)\left[\Sigma(x_i - \bar{x})^2\left(\frac{\delta u}{\delta x}\right)^2 + (y_i - \bar{y})^2\left(\frac{\delta u}{\delta y}\right)^2 + 2(x_i - \bar{x})(y_i - \bar{y})\left(\frac{\delta u}{\delta x}\right)\left(\frac{\delta u}{\delta y}\right)\right] \tag{7.11}$$

The first two terms are the variances σ_x^2 and σ_y^2, respectively:

$$\sigma_x^2 = \lim\left(\frac{1}{n}\right)\Sigma(x_i - \bar{x})^2 \text{ and } \sigma_y^2 = \lim\left(\frac{1}{n}\right)\Sigma(y_i - \bar{y})^2$$

The third term may be expressed in a form similar to the variance as:

$$\sigma_{xy}^2 = \lim\left(\frac{1}{n}\right)\Sigma(x_i - \bar{x})(y_i - \bar{y}) \tag{7.12}$$

This is called the covariance between the variables x and y. With these definitions, the approximate value of the variance of u can be written as:

$$\sigma_u^2 = \sigma_x^2 \left(\frac{\delta u}{\delta x}\right)^2 + \sigma_y^2 \left(\frac{\delta u}{\delta y}\right)^2 + 2\sigma_{xy}^2 \left(\frac{\delta u}{\delta x}\right)\left(\frac{\delta u}{\delta y}\right) \qquad (7.13)$$

This equation is known as the error propagation equation. The first two terms are averages of squares of deviations and, hence, always positive. The third term is the average of cross terms involving the products of deviations in x and y simultaneously. If the fluctuations in x and y are independent (uncorrelated), then on the average we would expect to find as many approximately equal negative values for this term as positive values, resulting in their net contribution to be very small or negligible relative to those for the first two terms for a large number of random observations. Then the above equation reduces to:

$$\sigma_u^2 = \sigma_x^2 \left(\frac{\delta u}{\delta x}\right)^2 + \sigma_y^2 \left(\frac{\delta u}{\delta y}\right)^2 \qquad (7.14)$$

If u is a function of only one variable, x:

$$\sigma_u = \sigma_x \left(\frac{\delta u}{\delta x}\right) \qquad (7.15)$$

If u is a function of more than two independent variables, the equation will include additional similar terms corresponding to the additional variables. It is of interest to use the above error propagation equation to simple addition/subtraction of x and y, as:

$$u = x \pm y$$

$$\sigma_u^2 = \sigma_x^2 + \sigma_y^2 \pm 2\sigma_{xy}^2 \qquad (7.16)$$

σ_u^2 can, in principle, vanish if the covariance σ_{xy}^2 has the proper magnitude and sign. This would happen if the fluctuations were completely correlated so that each erroneous observation of x would be exactly compensated by a corresponding erroneous observation in y. This brings out the complications in compounding errors, if the errors in the variables are correlated.

7.6 STANDARD DEVIATION OF THE MEAN OF *n* MEASUREMENTS

The mean of n measurements, $x_1, x_2, \ldots x_n$, each with the same standard deviation σ will be given by:

$$\bar{x} = \left(\frac{1}{n}\right)(x_1 + x_2 + x_3 + \cdots x_n) \qquad (7.17)$$

Partial derivatives of x with respect to each variable give (Zou, 2007):

$$\left(\frac{\delta \bar{x}}{\delta x_1}\right) = \left(\frac{\delta \bar{x}}{\delta x_2}\right) = \cdots = \left(\frac{\delta \bar{x}}{\delta x_n}\right) = \left(\frac{1}{n}\right)$$

Error Analysis

Substituting these in the error propagation equation:

$$\sigma_x^2 = \left[\left(\frac{\delta \bar{x}}{\delta x_1}\right)^2 \sigma^2 + \left(\frac{\delta \bar{x}}{\delta x_2}\right)^2 \sigma^2 + \cdots \left(\frac{\delta \bar{x}}{\delta x_n}\right)^2 \sigma^2\right]$$

$$= n\left(\frac{\sigma}{n}\right)^2 = \left(\frac{\sigma}{\sqrt{n}}\right)^2 \qquad (7.18)$$

If the n measurements are not equally precise, but have different standard deviations, $\sigma_1, \sigma_2, \ldots \sigma_n$, then each measurement is multiplied by a suitable weighting factor w_i (usually taken as the inverse square of the standard deviation) to give the weighted mean as:

$$\bar{x}_w = \frac{(w_1 x_1 + w_2 x_2 + \cdots w_n x_n)}{(w_1 + w_2 \cdots w_n)}$$

$$= \frac{\sum (w_i x_i)}{\sum (w_i)} \qquad (7.19)$$

As before, the standard deviation of the weighted mean can be calculated to be:

$$\sigma_{\bar{x}w} = \frac{1}{(\Sigma w_i)^{1/2}} \qquad (7.20)$$

For example: If the Uranium-Lead (U-Pb) ages of five homogeneous zircon crystals are respectively 1,105 ± 35 Ma, 1,195 ± 60 Ma, 1,125 ± 20 Ma, 1,150 ± 25 Ma, and 1,095 ± 45 Ma, the weighted mean age and its error will be 1,130 ± 13 Ma. Note that the age with the least uncertainty contributes dominantly to the final age and uncertainty results (Zou, 2007).

7.7 Joint Variation of Two or More Random Variables

Joint variation of two or more variables is far more informative than variation in a single variable in geochemical research. When only two variables are involved we speak of a simple correlation. When more than two variables are involved we speak of multiple correlations. We will consider only simple or bivariate correlation in the following.

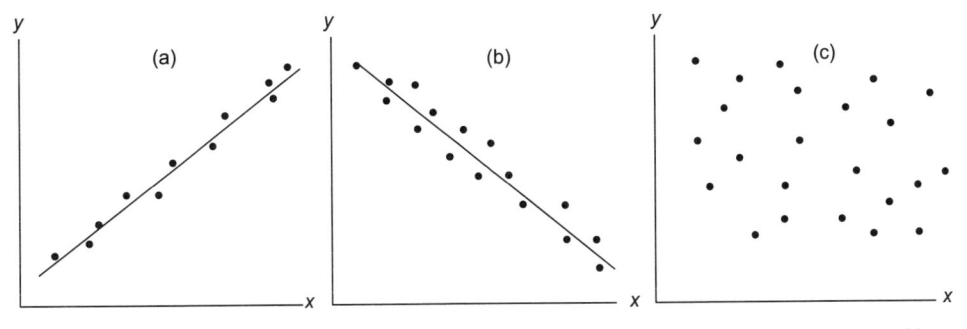

Figure 7.3 Some examples of correlation: good positive correlation (a), moderate negative correlation (b), and no correlation (c).

The degree and sense of association between two variables can be qualitatively seen on a plot of several pairs of the two variables (say, x and y), called the scatter diagram (Figure 7.3). If points in this diagram seem to lie close to a straight line, the correlation is called linear. If y tends to increase as x increases, as in Figure 7.3a, the correlation is said to be positive or direct. If y tends to decrease as x increases, as in Figure 7.3b, the correlation is called negative or inverse. If all points seem to lie near a curve, the correlation is called nonlinear. If there is no obvious relationship between the variables, as in Figure 7.3c, we say that there is no correlation or the variables are not correlated.

A quantitative measure of the mutual variability of a pair of variables can be calculated using the computational procedures used to calculate the variance of a single variable. In the previous section we introduced one such measure, called the covariance, which represents the joint variation of two variables about their respective means. Just as the variance measures the spread of values around a central point, the covariance measures the distribution of values of two variables about their respective means. It is defined as:

$$\sigma_{xy}^2 = \sum \frac{(x_i - \bar{x})(y_i - \bar{y})}{(n-1)}, \tag{7.21}$$

when there are n values of the variable x ($x_1, x_2 \ldots x_n$) and of variable y ($y_1, y_2, y_3 \ldots y_n$). Notice the similarity between equations 7.1 and 7.21. $(n - 1)$ is the degrees of freedom in a measurement of a limited number of pairs. The intensity of the relationship of the two variables is given by the linear correlation coefficient r, which is defined as covariance divided by the product of the standard deviations for x and y.

$$r = \frac{\text{Covariance}}{[(\text{variance }(x)\ \text{variance }(y)]^{0.5}}$$

$$= \frac{\sigma_{xy}^2}{\sigma_x \cdot \sigma_y} \tag{7.22}$$

Because the denominator is always positive, the algebraic sign of r is the same as that of σ_{xy}^2. The covariance of two variables may be equal, but cannot exceed the product of their standard deviations. So the linear correlation coefficient can range only from +1 to −1. $r = +1$ indicates a perfect direct relation between the two variables; $r = -1$ indicates a perfect inverse relation. Between these two extremes is a spectrum of less than perfect relationships including 0, which precludes any sort of linear relationship.

Some correlations between variables do not reflect any intrinsic relationship, but are induced by an operation or transformation that has been performed on the variables, for example, correlation induced between ratios of geochemical variables. The dangers of ratio correlation have been discussed by Rollinson (1993). Given a set of variables, $x_1, x_2, x_3 \ldots$ without any correlation between them, pairs of ratios formed from them, such as (x_1/x_2, x_3/x_2) (x_1/x_2, x_1/x_3) and (x_1, x_2/x_1) will be highly correlated. Rollinson cites the example of a very low correlation coefficient between Na_2O and K_2O concentrations in a felsic rock, but very high correlation (correlation coefficient ~0.88) when the two concentrations are ratioed to titanium (Ti) concentration. Plots of trace element ratios are also subject to spurious relationships. Isotopic ratios with a common denominator are quite common in radiogenic isotope geochemistry and, hence, some of them may show correlation over

Error Analysis

and above any intrinsic association between them. Perfect linear correlation of mixing curves has already been covered in Chapter 4.

7.8 REGRESSION ANALYSIS

In the case of most geochemical variables theoretical considerations do not predict any specific functional relationship (linear or other) between them. In fact, it may even not be clear if one variable is independent and the other dependent. If their correlation coefficient turns out to be very high ($r > 0.95$) a very strong linear relation is indicated. In contrast, theoretical considerations in radiogenic isotope geochemistry often predict a strict functional dependence between parent (as independent variable) and daughter (as dependent variable) isotope concentrations. The best example is the basic age equation which implies a perfect linear relation between the present daughter isotope composition (α) and parent/daughter ratio (μ), which is the basis of the isochron concept. The mathematical quantification of the relationship between geochemical variables is useful for predictive purposes, whereas, it is crucial for calculation of a common age and initial daughter ratio in a set of related natural objects from the slope and intercept, respectively, of the defined isochron.

This quantification is often carried out by fitting a straight line to a set of pairs of variables and finding the equation of that straight line ($y = a + bx$) in terms of its slope (b) and y-intercept (a). Many methods for fitting a straight line to data points have been used, the most common being the linear least squares regression. As the name implies, this method uses the intuitively reasonable approach that the best line is the one which minimizes the sum of squares of deviations (defined in some way) of data points from the line. The directions in which the deviations are to be measured in a particular case depend on the quality of the data points on hand.

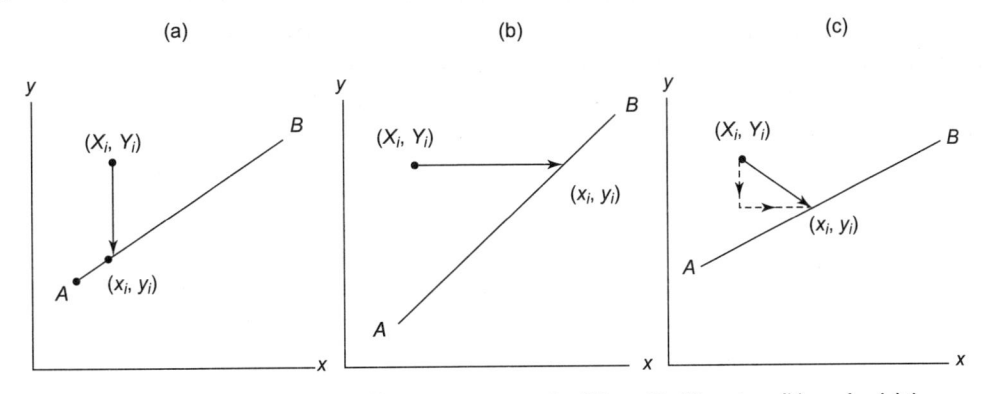

Figure 7.4 Schematic illustration of least-squares regression fitting with different conditions of weighting

Let (X_i, Y_i) be the observed points, and (x_i, y_i) the predicted values on the line, $y = a + bx$. Note that the observed or measured variables are now represented by (X_i, Y_i) and not (x_i, y_i), as have been so far. If the x variable is well defined (free of error) and only the y variable is subject to experimental errors, $X_i = x_i$, but $Y_i \neq y_i$. It is obvious that the sum (S) of the squares of vertical deviations (Figure 7.4a) from the fitted line is to be minimized. The sum is given by:

$$S = \Sigma(y_i - Y_i)^2 \tag{7.23}$$

If there are some reasons to trust some observations more than others, the sum to be minimized becomes:

$$S = \Sigma W(Y_i)(y_i - Y_i)^2 \qquad (7.24)$$

where $W(Y_i)$ is the weighting factor for the Y_i value. If, on the other hand, Y_is are free of error and all the experimental errors are in the X_is, $Y_i = y_i$ and $X_i \neq x_i$. The sum of the squares of the horizontal deviations will have to be minimized:

$$S = \Sigma W(X_i)(x_i - X_i)^2 \qquad (7.25)$$

where $W(X_i)$ is the weighting factor for X_i in the data set. The former is called regression of y on x and the latter regression of x on y (Figure 7.4b). If both X_i and Y_i are subject to measurement errors, as is inevitable in practice, $X_i \neq x_i$ and $Y_i \neq y_i$. The sum to be minimized will be squares of deviations in an oblique direction (Figure 7.4c)

$$S = \Sigma W(X_i)(x_i - X_i)^2 + W(Y_i)(y_i - Y_i)^2. \qquad (7.26)$$

The summation in each case above runs from $i = 1$ to n, where n is the number of paired observations used in the calculation. When only X_i or Y_i is subject to errors, the minimization of the sum (equations 7.24 or 7.25) is simple and straight forward, and the procedure, called the ordinary least squares regression, is most commonly used in physical and engineering sciences (Bevington, 1969). Even inexpensive electronic calculators are now programmed to carry out the minimization at the press of a button. It may appear that when both X_i and Y_i are subject to independent errors, minimization of the sum (Equation 7.26) is only marginally harder. Although Deming (1943) proposed the minimization of this sum to tackle two error regression as far back as 1943, it was only in 1966 that Derek York of the University of Toronto, proposed an exact and completely general solution to the problem of least squares fits with error on both coordinates. He was motivated to find the best slope and intercept of an isochron defined by imprecise data points, specifically in the context of geochronology. York's solution is widely used in geochronologic calculations, mainly due to the availability of many software packages to do the otherwise tedious calculation by hand (Ludwig, 2000). The full derivation of the York's technique is beyond the scope of this book. The following is only a simplified summary of the technique. Zou (2007) has given a complete derivation.

7.9 York's Solution

The problem is to find out the values of the parameters b (slope) and a (intercept) that yield the best straight line of the model equation:

$$y_i = a + bx_i \qquad (7.27)$$

where x_i and y_i are the true or adjusted values. The sum of squared deviations can, therefore, be rewritten as:

$$S = \Sigma (x_i - X_i)^2 W(X_i) + (a + bx_i - Y_i)^2 W(Y_i) \qquad (7.28)$$

To minimize S its partial derivatives $(\delta S/\delta x_i)$ $(\delta S/\delta a)$ and $(\delta S/\delta b)$, with respect to the three unknowns (x_i, a, b), are set equal to zero to get what are known as normal equations. The three normal equations are then solved for the three unknowns which will minimize S. The normal equation resulting from setting $(\delta S/\delta x_i) = 0$ leads to:

$$S = \Sigma W_i(a + bX_i - Y_i)^2$$

where

$$W_i = \frac{W(X_i)W(Y_i)}{[W(X_i) + b^2 W(Y_i)]} \quad (7.29)$$

W_i has the simple geometrical interpretation in terms of the equivalent variance in the y coordinate caused by the variance in the x coordinate. It may be treated as the overall or effective weight. The normal equation arising from setting $(\delta S/\delta a) = 0$ leads to:

$$a + b\bar{X} = \bar{Y} \quad (7.30)$$

where $\bar{X} = \Sigma W_i X_i / \Sigma W_i$ and $\bar{Y} = \Sigma W_i Y_i / \Sigma W_i$ are the weighted averages of X_i and Y_i in terms of W_i. The new expression for S is:

$$S = \Sigma W_i (bU_i - V_i)^2 \quad (7.31)$$

where $U_i = X_i - \bar{X}$ and $V_i = Y_i - \bar{Y}$.

Equation 7.30 shows that the fitted line will pass through the centroid point $(\bar{X}, \bar{Y})$ which represents the weighted centre of gravity of the data points. Finally, the normal equation from setting $(\delta S/\delta b) = 0$ leads to a cubic equation in b as:

$$pb^3 + qb^2 + rb + s = 0 \quad (7.32)$$

where

$$p = \Sigma \left[\frac{W_i^2 U_i^2}{W(X_i)} \right]$$

$$q = -2\Sigma \left[\frac{W_i^2 U_i V_i}{W(X_i)} \right]$$

$$r = \left[\Sigma W_i^2 U_i^2 - \Sigma \left(\frac{W_i^2 V_i^2}{W(X_i)} \right) \right]$$

$$s = \Sigma (W_i U_i V_i)$$

York (1966) termed Equation 7.32 the Least Squares Cubic (LSC, hereafter). This equation will have three roots, one of which should be the slope of the best line. Equation 7.32 is strictly not a cubic equation in b since the terms in W_i involve the unknown b. However, an approximate value of b can be used in W_i to get an exact cubic equation and solve it for b. This improved value for b is then used to calculate an improved W_i, and the cubic equation is re-solved. This iteration is continued until two consecutive estimates of b are within the required precision. Once the final b is obtained, the best intercept is given by:

$$a = \bar{Y} - b\bar{X} \quad (7.33)$$

York gave the variances of b and a with sufficient degree of accuracy as:

$$\sigma_b^2 = \left(\frac{1}{n-2} \right) \left[\frac{\Sigma W_i (bU_i - V_i)^2}{(W_i U_i^2)} \right] \quad (7.34)$$

$$\sigma_a^2 = \sigma_b^2 \left[\frac{\sum W_i X_i^2}{\sum W_i} \right] \qquad (7.35)$$

Williamson (1968) gave b as:

$$b = \left[\frac{\sum W_i V_i Z_i}{\sum U_i Z_i} \right] \qquad (7.36)$$

where $Z_i = W_i \left[\left(\frac{U_i}{W(Y_i)} \right) + \left(\frac{bV_i}{W(X_i)} \right) \right] \qquad (7.37)$

Williamson (1968) also gave exact expressions for the variances of the slope and intercept, which are far more complex than the approximate expressions above.

York (1967) has shown that the graphical representations of the somewhat complicated analytical expressions for the LSC have a simple geometric meaning; a few of them are as shown in Figure 7.4.

The two error regression given by York (1966) do not allow for any correlation between the errors of the two coordinates. As this correlation may be significant in some isochron plots, York (1969) gave a generalized version of the uncorrelated case. The summation to be minimized now becomes:

$$S = \sum R_i (a - bX_i - Y_i)^2 \qquad (7.38)$$

Where the effective weighting factor is:

$$R_i = \frac{W(X_i)W(Y_i)}{[W(X_i) + b^2 W(Y_i) - 2br_i \alpha_i]} \qquad (7.39)$$

where $\alpha_i = \sqrt{W(X_i)W(Y_i)}$. R_i differs from W_i in the uncorrelated case by a third term in the denominator containing the correlation coefficient r_i (not to be confused with the third coefficient of the cubic Equation 7.32) between experimental errors in X and Y for point i. The best slope is now given by a quadratic equation in b. York (1969) also gives an explicit expression for b as:

$$b = \frac{\left\{ \sum R_i^2 V_i \left[\frac{U_i}{W(Y_1)} + \left(\frac{bV_i}{W(X_i)} \right) \right] - \left(\frac{r_i V_i}{\alpha_i} \right) \right\}}{\left\{ R_i^2 U_i \left[\frac{U_i}{W(Y_i)} + \left(\frac{bV_i}{W(X_i)} \right) - \left(\frac{br_i U_i}{\alpha_i} \right) \right] \right\}} \qquad (7.40)$$

As the slope term b appears on both sides of this equation, it must be solved iteratively, as before. Variance of b and a are given with sufficient accuracy by:

$$\sigma_b^2 = \frac{1}{(\sum R_i U_i^2)} \qquad (7.41)$$

$$\sigma_a^2 = \sigma_b^2 \left(\frac{\sum R_i X_i^2}{\sum R_i} \right) \qquad (7.42)$$

The exact expressions for σ_b and σ_a are intimidating, as those given by Williamson (1968) in the uncorrelated case.

7.10 Measure of Goodness-of-fit

York (1966) assumed that the scatter of the data points from perfect collinearity is mainly due to experimental uncertainties, and proposed a well known statistical analysis to test this assumption. The minimized value of S has a chi-squared distribution with an expectation value of $n - 2$ for a large number n of observations (points). In other words, the ratio ($S/n - 2$) should, on an average, be close to unity if the scatter of data about the fitted line is consistent with experimental uncertainties. This ratio has been used as a goodness-of-fit parameter and as a valid measure of the quality of isochrons in geochronology. (McIntyre et al., 1966; Brooks et al., 1972). It is called the Mean Square Weighted Deviates (MSWD) (Harmer and Eglington, 1990). The other name, Mean Sum of Weighted Deviates misses the important qualification 'squared'. A MSWD substantially >1 would, therefore, imply that either: (1) the measurements are less precise than supposed or assigned or (2) the scatter in the data exceeds that can be accounted for by analytical error alone. The latter means that even if measurement errors are reduced to zero, the data points will still show a scatter. In practice, the number of measurements is small and the magnitude of analytical errors is not exactly known. So a limiting value of MSWD to qualify a fitted line as an isochron is about 2.5 (Brooks et.al., 1972). These authors proposed that regression fits with MSWD > 2.5 indicate scatter in excess of experimental errors and should be regarded as 'errorchrons' to distinguish them from legitimate isochrons. It has been suggested that a suitable increase (usually by the factor $\sqrt{MSWD}$) of the assigned experimental errors will convert an overt errorchron into a covert isochron, but with a correspondingly magnified error on slope and intercept. Instead of so hiding excess scatter as an increased error in reported age, it is preferable to suggest its possible cause during the geological history of the dated samples.

8 Meteorites: Link between Cosmo- and Geochemistry

> It is as if Nature had taken a sample at each step of the planetary formation process, had kept it intact somewhere for 4.5 billion years, and had finally sent it to us from the sky so that we could study it in our laboratories
>
> C J Allegre

8.1 INTRODUCTION

From the point of view of a present-day observer in the Solar System, the time scale of the universe can be split into two major subdivisions—presolar and solar—as shown schematically in Figure 8.1. The presolar period begins with the big bang at time t_o (as measured from the present) and ends with the isolation from the Interstellar Medium (ISM) at time t_i of a small portion of the dust and gas in the ISM to form the Solar System. The immediately following solar era begins at time t_i with the formation of the solar nebula as a separate entity, and ends at the present time. Within these two broad time zones, one can distinguish discrete events and protracted processes even if some of the processes are yet unfamiliar. For example, the presolar era includes the cosmological or primordial synthesis of hydrogen (H) and helium (He) immediately after the Big Bang, formation of galaxies, synthesis of heavier elements from H and He in the interior of stars, and its termination in the material destined to form the Solar System. The solar era comprises the collapse of the isolated gas cloud into a central massive body (the potential sun) and a thin disc around the central object, condensation of the material in the disc into chemical compounds at t_c, aggregation and accretion of the chemical species into the planets and other objects of the solar system at t_a, and subsequent internal evolution of individual planets to the present time.

As geologists recognized the immensity of geologic and, hence, the solar time from observations of large scale geologic structures and processes, astronomers and astrophysicists relied on observations of large scale structures in the universe, like stars, star clusters, and galaxies to infer an even longer presolar time scale. The best known of such indirect dating methods is the mutual recession of galaxies, as evidenced by their spectral red shift. The most precise result, so far, for the Big Bang, as based on the rate of expansion of the universe, is 13.4 ±1.6 Ga (Lineweaver, 1999).

Subtraction of the more precisely determined solar time of 4.6 Ga gives the duration of the presolar period as ~9 by.

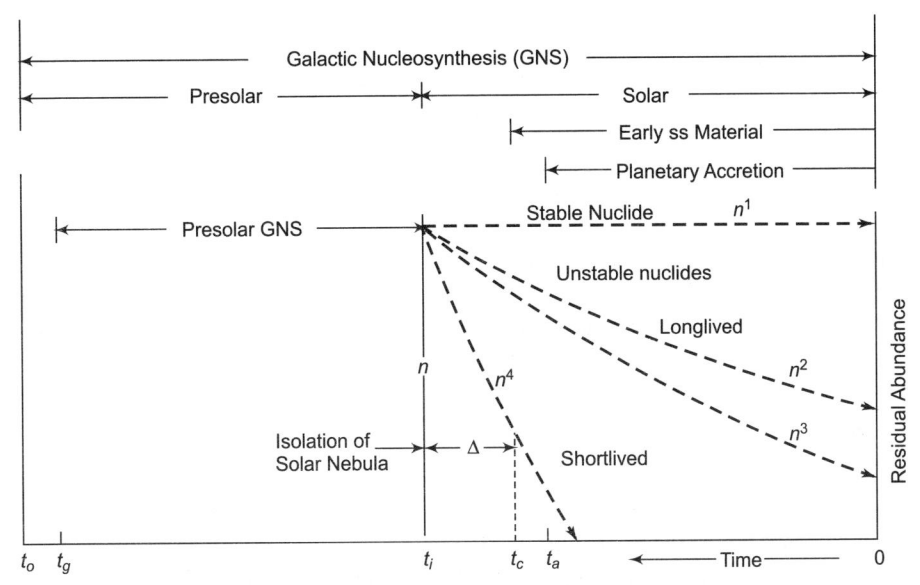

Figure 8.1 Schematic diagram showing the time scale measured from the present to the origin of the universe. Galactic Nucleosynthesis (GNS) continues over the entire evolution of the galaxy. t_i is the time of isolation of the solar nebular material from galactic nucleosynthesis, t_c the time of condensation of the solar nebula, and t_a the time of accretion of nebular condensates

8.2 Nucleocosmochronology

We noted in Chapter 3 that all the elements in the Solar System must have been formed prior to its own formation as a distinct entity at ~ 4.6 Ga. This raises the question of the time of formation of the elements that comprise the Solar System. We also noted that light elements up to iron (Fe) were synthesized by fusion reactions in the interior of stars, and elements heavier than Fe were synthesized by neutron-capture reactions at a slower rate (*s*-process) in massive stars and at a much faster rate (*r*-process) during their terminal explosion, collectively known as GNS. These explosions also served to return synthesized elements to the ISM to be recycled into the next generation of stars. Because of the generally higher heavy element content of younger stars relative to older stars, GNS must be a protracted process involving the births and deaths of many generations of stars in our galaxy. In the schematic diagram (Figure 8.1) the duration of the GNS for the Solar System material is shown to be between t_g and t_i. The variation in the abundance of different nuclear species during this period is an important astrophysical problem that we shall not go into in this book. We assume that the normalized abundance of any nuclear species in the solar nebula, just after its isolation, is n_i. If the species is stable, its abundance will be conserved at its initial value n_i^1, as shown by the horizontal dashed line:

$$n_i^1 = k^1 \cdot (t_g - t_i) = n^1 \tag{8.1}$$

where k^1 is the average production rate and $(t_g - t_i)$ the duration of production. The abundance n_i^2 of a radioactive species assumed to be in approximate steady state (production rate is ≈ decay rate) will be:

$$n_i^2 = \frac{k^2}{\lambda^2} \tag{8.2}$$

where k^2 and λ^2 are the production rate and decay constant, respectively, of the species. If the half-life of the species $(0.693/\lambda^2)$ is comparable to the time scale of the solar era, its initial abundance will exponentially decrease to the present abundance n^2, as shown by the dashed curve, n^2:

$$n^2 = \left(\frac{k^2}{\lambda^2}\right) \exp(-\lambda^2 t_i) \tag{8.3}$$

where t_i is the decay interval. Rutherford (1929) suggested that this time dependence could be used to estimate the mean age of this species. The need to measure absolute abundances was eliminated by taking the ratio of the residual abundances of two related nuclides (n^2 and n^3, respectively) with long, but comparable decay constants, λ^2 and λ^3, as:

$$\left(\frac{n^2}{n^3}\right) = \left(\frac{k^2}{\lambda^2}\right)\left(\frac{\lambda^3}{k^3}\right) \exp[-(\lambda^2 - \lambda^3)t_i] \tag{8.4}$$

Rutherford used the two isotopes of uranium (U), ^{238}U and ^{235}U in calculating their mean age. The large uncertainty in the theoretical production rates (k^2 and k^3) precluded a precise calculation. The pair of nuclides is called a chronometer pair. The science of determining mean ages and, hence, the presolar time scale is called nucleocosmochronology (Fowler, 1972). Considerable progress has been made since the maiden work of Rutherford to calculate the mean age of many chronometric pairs like ^{232}Th-^{238}U and ^{187}Re-^{187}O. In particular, Schramm and Wasserburg (1970) have proposed a method of calculating the mean age of chronometer pairs that is essentially independent of the model assumed for nucleosynthesis. A simplified treatment of this type of calculations has been given by Schramm (1974a, b).

8.3 Extinct Nuclides and Formation Interval

Heavy elements (nuclides) like uranium and thorium (Th) are created by rapid neutron absorption (*r*-process) by lighter seed nuclides during brief supernova explosions of massive stars. For this reason this process is also called explosive nucleosynthesis. All nuclei generated in the neutron hail are very neutron-rich and far away from the band of stability (Chapter 1). The *r*-process nuclides lighter than bismuth (Bi) reach the nuclear stability zone by successive beta (β) decays, and nuclides heavier than Bi reach stability by both alpha (α) and (β) decays. Whereas, U and Th nuclides produced in the *r*-process have sufficiently long half-lives to survive to the present (extant nuclides); many other *r*-process nuclides have much shorter half-lives compared to the length of Solar System history. Thus, they have all decayed to extremely low, usually undetectable, concentrations by the present time. Such short-lived nuclides are referred to as 'extinct nuclides' as opposed to extant nuclides like ^{238}U. Their former presence can be inferred from an unambiguous anomaly in the isotopic abundance of their stable daughter isotope (Chapter 3). Extinct nuclides with half-lives in the range 10^6 to 10^8 years have attracted considerable astrophysical interest, as these half-lives are comparable to the free fall time of the isolated solar nebula on dynamical grounds (Wetherill, 1975), and, hence, can be used to estimate the time interval between the isolation of the solar nebula from the ISM and its condensation into the earliest Solar System material. A selected list of such nuclides is given Table 8.1.

Table 8.1 Selected List of Short-lived (extinct) Nuclides

Parent	Daughter	Half-life (10^6 y)
^{10}Be	^{10}B	1.5
^{26}Al	^{26}Mg	0.7
^{53}Mn	^{53}Cr	3.4
^{60}Fe	^{60}Ni	1.45
^{107}Pd	^{107}Ag	6.5
^{129}I	^{129}Xe	17.3
^{146}Sm	^{142}Nd	101
^{182}Hf	^{182}W	9.0
^{205}Pb	^{205}Tl	15.0
^{244}Pu	^{132,134,136}Xe	83.9

The steady state abundance of a short-lived nuclide (like n^4) at time t_i:

$$n_i^4 = \frac{k^4}{\lambda^4} \qquad (8.5)$$

where k^4 is its steady production rate and λ^4 its decay constant (much larger than that of an extant nuclide like ^{238}U). The abundance of this nuclide, relative to that of any of its stable counterpart (say n^1), is:

$$\left(\frac{n_i^4}{n_i^1}\right) = \left(\frac{k^4}{\lambda^4}\right) [k^1(t_g - t_i)]^{-1} \qquad (8.6)$$

If the production rates k^1 and k^4 are roughly equal, the relative abundance simplifies to:

$$\left(\frac{n_i^4}{n_i^1}\right) = [\lambda^4 \cdot (t_g - t_i)]^{-1} \qquad (8.7)$$

Assuming production of the stable nuclide n^1 over the age of the galaxy at ~10^{10} years and a typical mean life ($1/\lambda^4$) of 10^7 years for the unstable nuclide n^4, their relative abundance at the time of isolation comes out as small as ~ 10^{-3}. (n_i^4/n_i^1) can be significantly higher than the steady state value only if there was a 'late spike' of explosive nucleosynthesis immediately prior to the formation of the Solar System. The relative abundance at the time of condensation t_c of the nebular cloud will be:

$$\left(\frac{n_c^4}{n_c^1}\right) = \left(\frac{n_i^4}{n_i^1}\right) \exp[-\lambda^4(t_i - t_c)] = \left(\frac{n_i^4}{n_i^1}\right) \exp(-\lambda^4 \Delta) \qquad (8.8)$$

where Δ is the time interval between the isolation of the solar nebula and its condensation, as shown. Δ is called the formation interval by some workers (Wetherill, 1975). Thus, the relative change in the abundance of two nuclides as a consequence of their different half-lives during the formation interval permits a calculation of the duration of this period. The accuracy of the absolute value of this interval depends on the actual initial value and, hence, on details of galactic nucleosynthesis.

The actual experimental determination of formation intervals is considerably more complicated than may be apparent from the foregoing simple expressions. It is contingent on a favourable combination of many stringent conditions, as (1) Substantial, preferably more than the steady state, abundance of the chosen short-lived nuclide in the freshly separated solar nebula; (2) Residual persistence of the nuclide at the end of the formation interval; (3) Its preferential incorporation at measurable levels in some of the condensed phases; (4) Parallel depletion of the daughter element in these phases, so as not to drown or dilute the feeble radiogenic daughter of the extinct nuclide; (5) Preservation of such phases in pristine condition to the present time; and (6) Availability or accessibility of such phases for laboratory analysis. An additional, but implicit requirement is our current analytical capability. We will defer the experimental results on formation intervals based on a number of extinct nuclides to the next chapter on early Solar System chronologies, and focus on the far greater significance of Solar System samples, if they meet at least the last two conditions above.

8.4 Meteorites

We noted in the first chapter that the elemental composition of meteorites, especially the primitive carbonaceous chondrites, matches that of the sun and, hence, the solar nebula, except for very volatile elements. This raised the real possibility that such meteorites could represent tangible chemical species and components condensing from a cooling cloud of solar composition, and the raw material of the planets and other smaller bodies of the Solar System. This realization spurred extensive studies of the mineralogy, texture, isotope composition, and formation ages of meteorites of various types (McSween and Huss, 2010).

These studies have convincingly shown that meteorites are the oldest known objects in the Solar System that are available for direct study in the laboratory with the exception of some electromagnetic radiation and starlight (Wasserburg, 1987). Meteorites are, therefore, the closest link between astrophysical processes just before isolation of the solar nebula and pre-planetary processes in the early Solar System, as stated succinctly by Allegre in the beginning of this chapter. Wetherill (1975) has reviewed how radiogenic isotope studies on meteorites have advanced our knowledge of major events in the earliest history of the Solar System. The literature on meteorites is highly specialized and extensive, and, hence, is beyond the scope of this book. The reader is referred to the books by Wasson (1974, 1985). Given below is a concise overview of meteorites to help uninitiated readers follow meteoritic data in the following chapters.

Pieces of stone and, occasionally, iron falling from the sky accompanied by spectacular light and/ or sound effects must have puzzled men, even in prehistoric times, as some divine phenomena with mysterious significance. Ancient Greek, Roman, Egyptian, Chinese, and Indian literature record such events. In historic times, such objects must have been destroyed for superstitious reasons. Kings and Eskimo hunters have forged large chunks of iron meteorites into swords and arrowheads (Dalrymple, 1991). The earliest observed fall, recovery, and preservation of a meteorite is the 127 Kg stone meteorite that fell on 16 November 1492, in the village of Ensisheim in France. Meteorites are named after the place where they are seen to fall (falls) or found by chance (finds). Despite sighting numerous falls from the sky at great speeds, scientists were skeptical about their extra-terrestrial origin. The United States (US) president, Thomas Jefferson, sarcastically remarked in 1807, 'It is easier to believe that Yankee professors (from the Yale University) would lie than that stones would fall from the sky'. A few years earlier in 1794, E. Chladni, a German physicist of acoustics fame described the 160 Kg Pallas Fe meteorite found in Siberia, and the 15,500 Kg Otumpa iron meteorite in Argentina, and concluded, after rejecting other possibilities, that they were of cosmic origin. Later, he extended this conclusion to stone meteorites also.

Meteorites are now known to be extraterrestrial objects moving in heliocentric orbits that intersect the orbit of the earth, and possess sufficient mechanical strength and bulk to survive supersonic passage through the atmosphere, unlike meteors. Atmospheric ablation and alteration is confined to a thin surface layer, and the interior of the meteorite is in the same physical, chemical, and mineralogical state as it was in interplanetary space. Meteorites are most certainly fragments from small planetary bodies referred to, in general, as parent bodies.

8.4.1 Classification of meteorites

Meteorites are now classified into two broad categories: undifferentiated and differentiated. The former are so named because their chemical composition is close to the solar composition, except for very volatile elements. These are also called chondrites, because most of them contain ~1mm diametre spherical objects, called chondrules. Chondrites have been subdivided into carbonaceous (C), ordinary (O), and enstatite (E) chondrites. Table 8.2 shows further subdivisions of these groups of undifferentiated meteorites. Carbonaceous chondrites, particularly those designated as Cl, contain carbon and very volatile elements, and are closest to the solar composition. They are, therefore, regarded as the most primitive meteorites, that is, closest to the original composition of the solar nebula. The numerically abundant ordinary chondrites are divided into L, LL, and H groups according to their iron contents. Enstatite chondrites contain enstatite ($MgSiO_3$) as the main silicate phase. Most of these chondrites have undergone various degrees of thermal metamorphism on their parent bodies, resulting in a number of changes. Thus, the chemical division of chondrites into C, LL, L, H, and E classes is further classified into various petrologic types on a scale from 1 to 7. Textural and mineralogical changes from 1 to 7 are evident from the increasing homogeneity or equilibration of minerals, decreasing volatile content, obliteration of primary textures, and the blurring contrast between chondrules and matrix is attributed to increasing degree of heating (400 to 1,000° C) of original material (Van Schmus and Wood, 1967). For example, least equilibrated petrologic types 1, 2, and 3 are found only among the chemically least-processed carbonaceous chondrites, whereas, types 4 to 7 occur among ordinary and enstatite chondrites. According to this combined chemical-petrologic classification, H3 refers to a relatively unmetamorphosed H chondrite, and E7 to a highly metamorphosed and recrystallized enstatite chondrite.

Differentiated meteorites include: (1) irons consisting mainly of Ni-Fe alloy; (2) achondrites, consisting mainly of silicates; and (3) Stony iron, consisting of metal and silicate in similar amounts. While some achondrites resemble basalts on the earth, irons resemble the core of the earth. This implies metal-silicate fractionation and segregation due to melting in the parent bodies of differentiated meteorites. Consequently, the chemical composition and structure of differentiated meteorites do not represent the primary Solar System features. The subdivisions of achondrites, irons and stony-irons are not shown in Table 8.2. Differentiated meteorites, like achondrites and irons may be regarded as products of actual melting and chemical reconstitution of originally chondrite-like material.

Table 8.2 Classification of Meteorites

Type	Class	Subdivisions
Undifferentiated		
(Chondrites)	Carbonaceous (C)	CI, CM, CO, CV, CR, CH, CK
	Ordinary (O)	L, LL, H
	Enstatite (E)	EH, EL
Differentiated	Irons	Magmatic and Non-magmatic
	Stony Irons	Pallasites and Mesosiderites
	Achondrites	Ca-rich and Ca-poor

Only ~ 2,500 meteorites were known until 1969. Subsequently, the search for meteorites that fell over millions of years over the desolate Antarctic ice fields and deserts yielded more than 7,500 meteorites of all types. Though meteorites recovered from Antarctica and African deserts are finds, they have been preserved in their original condition because of the favourable environment in these regions. Staggering number of meteorites must have fallen in the oceans. The ~900 falls, recovered in the last few hundred years may not be representative of their parent bodies in the asteroidal belt, because all known meteorite falls were observed very recently in earth history, and so may not represent the long-term average. A few of the achondrites show convincing evidence of having been derived from the moon and Mars.

8.4.2 Main components of chondrites

Chondrites consist of varying proportions of four main components: Chondrules, Ca-Al Inclusions (CAIs), Amoeboid Olivine Aggregates (AOAs), and Matrix.

Chondrules, as mentioned earlier, are spheroidal ultramafic melt droplets. They are the principal constituents of many chondritic meteorites. They are made up of silicate, metal, sulfide, and glass phases. Chondrules show a wide variety in chemical and mineral compositions. While chondrule formation implies flash heating to melting temperatures of preexisting materials, the source of flash heating is still debated.

The CAIs contain aluminium (Al) and calcium (Ca) rich refractory minerals such as melilite, spinel, pyroxene, plagioclase, and perovskite. They are commonly found in carbonaceous chondrites, and more rarely in ordinary and enstatite chondrites. They are believed to have formed in a high temperature part (> 1,700 K) of the solar nebula.

The AOAs are irregular objects up to ~1 millimetre long. They comprise granular olivine, intergrown with the refractory phases diopside, anorthite, and spinel, and contain grains of Fe-Ni metal. The presence of olivine indicates that these aggregates are formed in the solar nebula at a lower temperature than CAIs. Close similarity of oxygen (O) isotopic composition with distinct ^{16}O enrichment between CAIs and AOAs implies that AOAs and CAIs are formed from the same parcel of gas in the solar nebula, but at different temperature intervals.

Matrix refers to dark, FeO–, and volatile rich material that is very fine grained (~1 micrometre). Primary constituents are Fe– and Ca poor pyroxene and olivine, and amorphous material, but magnetite, Fe metal, and a wide variety of silicates, sulfides, carbonates, and other minerals are also present. In carbonaceous chondrites, carbonaceous material is present in substantial quantities. On the whole, the composition of the matrix is complementary to that of chondrules. Very significantly, matrix includes grains of SiC, graphite, diamond, and other phases of anolmalous isotopic compositions. These 'presolar grains' are of great significance to nuclear astrophysics (Section 8.7.1).

8.5 Nebular condensation

Given a gas of solar composition cooling from a very hot condition it is straightforward to calculate the sequence of condensed phases as a function of temperature using standard thermochemical data. Grossman (1972) was the first to use thermodynamic equilibria to calculate the composition of phases in equilibrium with a gas of solar element concentrations at a pressure of 10^{-4} atmosphere and as a function of temperature. This scheme has subsequently been developed by others (Lewis, 2004). A simplified sequence (Lewis, 2004) of phases is given in Table 8.3.

Table 8.3 Solar Condensation Sequence at a Nebular Pressure of 10^{-4} bar

Temperature, K	Condensate
>1700	Refractory elements like W, Os, Ir, and Re
<1700	Refractory oxides of Al, Ca, and Ti (corundum, spinel, perovskite) with REEs, U and Th
ca 1450	Fe-Ni metal with Co, Cu, Au, Pt, Ag, and nonmetals like P, N, C
ca 1400	Magnesium silicates (Olivine, Mg-pyroxene)
ca 1020	Alkaline and alkali feldspars
ca 670	Iron sulfide, -oxide with Zn. Pb and As
ca 430	Hydrated silicates (amphibolite, serpentine, and chlorite)
ca 140	Ices (water, ammonia, methane, and rare gases)

The calculation assumes highly simplified conditions, and, hence, is not expected to correspond to the earliest Solar System condensates in an obvious way. But a convincing support for the condensation model came from a two-ton carbonaceous chondrite 'Allende' that fell in Mexico in February 1969 (a few months before the recovery of the very first batch of moon samples). Eight per cent of this now famous meteorite consists of tiny white Ca- and Al-rich inclusions (CAIs) composed of perovskite, melilite, and spinel predicted as the first refractory oxide condensates from the solar nebula. Even more strikingly, these inclusions contain tiny metal nuggets of refractory trace metals [rhenium (Re), molybdenum (Mo), tungsten (W) and platinum (Pt) group elements] surrounded by free iron particles and then by olivine and pyroxenes exactly as in the sequence. As we will see later, these CAIs have given the oldest ages. Further, the matrix of Allende, in which the CAIs are embedded, contains hydrous and other minerals that condense at low temperatures.

8.6 PLANETARY ACCRETION

The condensation of a solar nebula results in a protoplanetary disc laden with micron-size mineral grains. Aggregation of such grains into a few huge planetary objects presents a formidable problem, as any model of planetary formation must explain the following features of the Solar System:
1. Coplanar orientation and spacing of planetary orbits;
2. Progression of small inner planets to giant outer planets;
3. Angular momentum of the Solar System principally in the planets;
4. Compositional differences between dense inner planets, and light outer gaseous and icy planets;
5. Existence of satellites;
6. Existence of asteroids as a source of meteorites; and
7. Large impact craters on most solid planetary and satellite surfaces.

The standard model of planetary formation (Wetherill, 1990) is based on numerical simulations and envisages the following stages:
1. Coagulation of micron-sized mineral grains into (1–10 km, 10^9–10^{12} kg) planetesimals within 10^5 years through weak electrostatic, non-gravitational forces.
2. Amalgamation of the numerous planetesimals into a few tens of (~10^3 km, 10^{23}–10^{24} Kg) planetary embryos from their mutually constructive collisions followed by runaway growth of bigger bodies within 10^6 years.
3. Catastrophic collisions of the embryos resulting in the incorporation of disrupted embryos into a few large planets over 10^7–10^8 years. Such giant impacts could have led to large scale melting of the larger bodies.
4. Giant planets like Jupiter are believed to have a rocky core, a thick aqueous outer layer and possibly a middle icy layer. Models suggest that Jupiter first accumulated as a massive rocky embryo (~10 earth masses) and later captured two earth masses of solar nebular gas and also ices. The massive Jupiter is believed to have gravitationally prevented the growth of planetesimals in the asteroid belt into a single planet, and even stunted the growth of Mars to less than earth size. As the protosun heated up and entered the so-called T-Tauri stage, it emitted an intense stream of charged particles—the solar wind—radially along the nebular disk. This drove away any still uncondensed gas and dust, ending any further planetary growth. Recent observations of extrasolar Jupiter-like planets have led to an alternative model of for their formation.

Although giant impacts were based on numerical simulations, there are ample evidences in the Solar System for impacts of the scale needed.
1. The rotation axes of Venus and Uranus have been knocked out of alignment with those of the other planets.
2. Mercury has been stripped out of its outer crust.
3. Mars is distinctly asymmetric—its southern hemisphere features high mountains and ancient impact craters, whereas, its northern hemisphere has been scoured flat and covered with massive lava flows that followed a giant impact.

8.7 Isotope Abundances in the Solar Nebula

The direct products of stellar nuclear reactions are nuclides and not elements. Different types of stars will produce different mixtures of nuclides. Thus, the isotope abundances in the newly formed solar nebula will be a grand average of many presolar components and are far more informative than its chemical composition, for they give clues to dominant stellar nuclear reactions. These abundances are a primordial reference to the subsequent isotopic evolution in earth and other planets. They help recognition of any exotic presolar component. Astronomical observations indicate a dust content of ~1 per cent in interstellar clouds, presumably formed in red giant stars in the AGB phase. Any dust component in the solar nebula is believed to have been fully vapourized and, hence, isotopically homogenized before condensation of the Solar System material. This belief was validated by the close agreement of isotope ratios in a wide variety of terrestrial and extraterrestrial materials, except for the effects of the two well-understood processes: radioactive decay and mass dependent fractionation. Two spectacular findings since the 1970s reveal isotopic heterogeneities, on two widely different scales, in the solar nebula. These are discussed briefly in the following sections.

8.7.1 Presolar grains

The presence of presolar material with anomalous noble gas isotope signatures was noticed in trace acid residues of primitive meteorites (McSween and Guss, 2010). The carriers of such noble gases were identified as extremely small (~1 micron) grains. Grains identified so far are graphite spherules, nano-diamonds, silicon and other metal carbides, silicon nitride, and silicates like corundum, spinel, and hibonite. The unusual isotopic composition of the solid elements in the grains implies their genesis in a different stellar environment, outside and before the Solar System. This ancient 'star dust' provides exciting new insights into the dynamics of supernovae, the age and chemical evolution of the galaxy, nuclear physics, and processes in the outer envelopes of bygone stars. The presence of a few extrasolar microscopic grains in primitive meteorites does not necessarily imply an isotopically inhomogeneous solar nebula. They could have been trapped from an influx of stellar dust particles in meteorites during the accretion of their parent bodies (Chapter 9). A highly readable account of presolar grains or ancient stardust is given by Bernatowicz and Walker (1997). Streams of intestellar dust currently intersecting the Solar System have been reported. Comparison of the isotopic properties of these grains [Interplanetary Dust Particles (IDPs)] with those embedded in primitive meteorites promises to be highly rewarding.

8.7.2 Variations of oxygen isotopes

Oxygen is a major element and has three isotopes, ^{16}O, ^{17}O, and ^{18}O with widely different abundances: 99.76, 0.04, and 0.2 per cent, respectively. Variations in $^{18}O/^{16}O$ and $^{17}O/^{16}O$ ratios due to natural physical and chemical processes are extremely small, and, hence, are measured as fractional deviation in parts per thousand: - $\delta^{18}O$ and $\delta^{17}O$ relative to $^{18}O/^{16}O$ and $^{17}O/^{16}O$ ratios of a standard [Standard Mean Ocean Water (SMOW)]. O isotope fractionations caused by natural, and physical and chemical processes are mass dependent. That is, variations in $\delta^{18}O$ (mass difference of two units) are twice as much in $\delta^{17}O$ (unit mass difference) and will define a fractionation line of slope close to 0.5 on a plot of $\delta^{17}O$ vs $\delta^{18}O$. Only $\delta^{18}O$ is usually measured in terrestrial samples, as $\delta^{17}O$ measurements are more difficult due to the still lower abundance of ^{17}O. When R. Clayton of the University of Chicago measured both ratios in various terrestrial samples, lunar rocks, and meteorite material, he found that terrestrial and lunar samples did indeed plot on a line of slope 0.5, called the terrestrial fractionation line (Figure 8.2), whereas, meteoritic materials plot not on this line, but on a steeper line of slope 1. That means the variations of O isotopes in the latter cases are mass-independent (Clayton et al., 1993). This unexpected result was initially explained as due to the incorporation of almost pure ^{16}O produced by nucleosynthesis in some ancestral star.

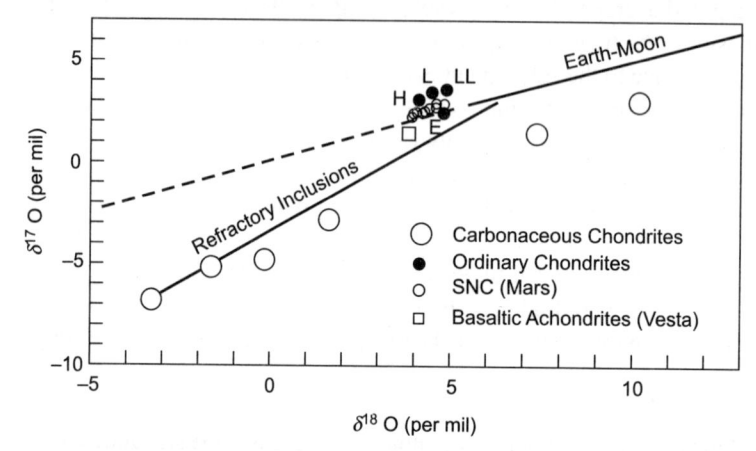

Figure 8.2 Isotopic composition of oxygen in different planetary bodies. LL, L, H, and E refer to ordinary chondrites and do not lie exactly on the earth-moon fractionation line of slope 0.5. Ca-Al rich inclusions in primitive carbonaceous chondrites fall on a line of slope one (Clayton, et al., 1991 as modified by Albarede, 2003).

Thiemens and Heidenrich (1983) proposed an alternative mechanism to produce Mass Independent Fractionation (MIF) that is based on the effects of molecular symmetry on the kinetics of formation of diatomic molecules. He demonstrated that ozone formed in an electrical discharge in diatomic O was equally enriched in ^{18}O and ^{17}O-relative ^{16}O. The asymmetric molecules ^{18}O^{16}O and ^{17}O^{16}O could be equally favoured over the symmetric molecule ^{16}O^{16}O in the solar nebula. Although the mechanism is not well understood, MIF does occur in nature and is documented in sulfates and sulfides. No presolar grain has so far revealed pure ^{16}O. Clayton himself now favours a photochemical mechanism (Clayton, 2002). Mass-independent variations in oxygen isotopes have been reviewed by Thiemens (2000).

9 Chronology of Meteorite History

> The problem then is the identification, within that piece of matter, those observations which can establish the nature and chronology of events whose memory has persisted. Such an approach requires the use of concepts, observations and measurement techniques from rather diverse fields and together define a new field, that of cosmochemistry or cosmophysics
>
> G J Wasserburg

9.1 Introduction

The last chapter contained a brief description of the processes leading up to the formation and subsequent changes of meteorites in the early Solar System. Figure 9.1 recapitulates the same, and, in addition, includes the later stages or events in the history of meteorites terminating with their capture by the earth.

A plausible list of distinguishable stages/events is:

1. Separation of a parcel of gas and dust from the interstellar medium, and its collapse into the proto-sun and proto-planetary nebular disc.
2. Formation from the cooling solar nebular disc of the precursor materials of chondrites, like CAIs, AOAs, low temperature compounds, and finally chondrules in a cool environment.
3. Accretion of mineral grains in various proportions, and chondrules into km-sized planetesimals (parent bodies of meteorites) including presolar grains directly from the interstellar medium
4. Metamorphism, fluid alteration, recrystallization on a local scale on the surface of, and even melting inside of parent bodies that escaped further growth in size into embryos and eventually planets.
5. Impact brecciation and shock reheating, presumably a few times.
6. Collisional ejection of metre-sized fragments (meteorites) from parent bodies, and their exposure to galactic cosmic rays in space.

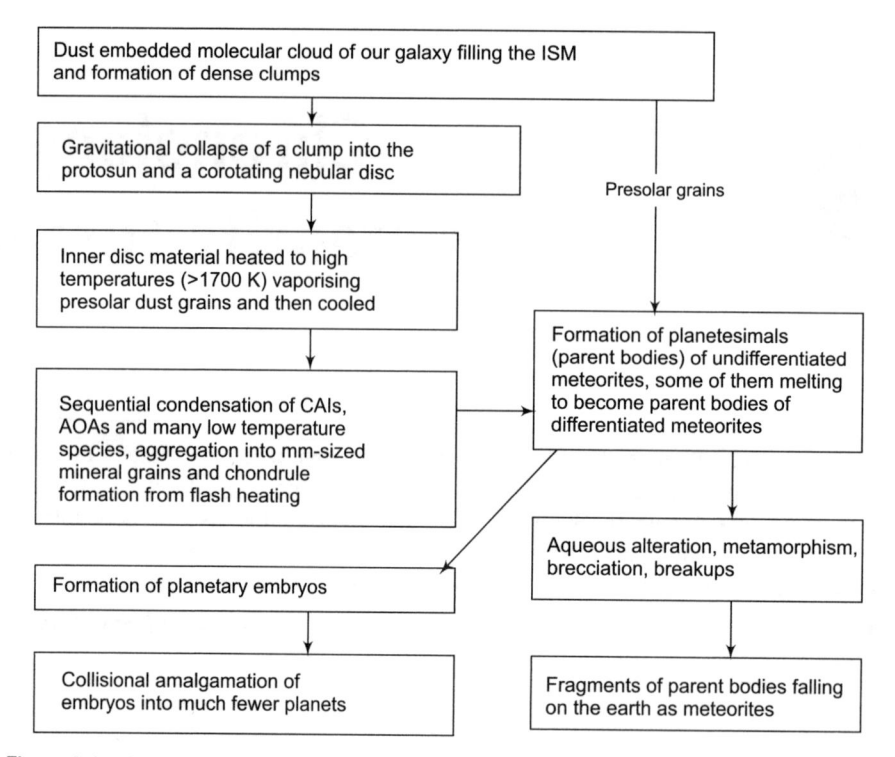

Figure 9.1 Schematic diagram of processes involved in the formation of meteorites and their components

7. Capture of meteorites by the earth, and their recovery later by chance (Finds).
8. Capture of meteorites by the earth and their prompt recovery (Falls).

This sequence of events is, of course, very general and meant as a useful framework for discussion. For example, not all stages may be recorded in a single meteorite. Some processes, like metamorphism, may be missing in primitive carbonaceous chondrites. Some events may have overlapped in time. The relative order of at least some of the events is obvious. For example, a carbonaceous chondrite is a nonequilibrium assemblage of refractory inclusions, metal, sulfide, and matrix, each of which formed independently. So, the chondrite should postdate these constituents. A long-range goal of meteorite research is to quantitatively date every stage precisely and accurately. In fact, one of the major areas of current research in cosmochemistry is to precisely date the closely-spaced events in the very early Solar System—the first four stages, in particular (McSween and Huss, 2010). It has become very clear now that the meteorite record of nebular condensation and accretion is very complex. We will restrict ourselves in this book only to a broad outline of the results on early Solar System evolution.

It is useful at this stage to distinguish between the terms, date, and age. A date is a number, with dimensions of time, calculated using measured isotopic ratios and the decay equation. It is the outcome of a careful analytical exercise. A date becomes an age when the former is interpreted as the time of occurrence of a particular event, or process. Accurate interpretation of a date into an age is not easy, given the complicated nature of embryonic processes in the Solar System. It is also useful to note the subtle differences in the meanings of units of time used in the literature, and in this book. Any period of time over which something happened will be given in units of Ky, My and Gy (for kiloyears, megayears, and gigayears, respectively). Time of occurrence of events, as measured from the present with the conventional meaning of age, will be given as Ka (kilo-annum), Ma (mega-annum), and Ga (giga-annum), respectively. Radiometric ages will be given in these units.

Chapter 5 describes many ways in which radioactive isotopes and their stable and unstable daughter isotopes can be used for radiometric dating of natural processes. It must be recalled that the one common feature of all these different methods is that they all measure only a radioactive decay interval. If a natural process fractionates a parent element from its daughter element, the subsequent decay of the parent isotope will date the process. If the process fractionates the parent element very strongly against the daughter element, the process can be dated with high time resolution from a single sample with negligible correction for any small initial presence of daughter. For a given parent-daughter ratio in a sample, the larger the decay constant, the higher is the time resolution. If the fractionation between the elements is only partial, time resolution will be lower, and also the initial daughter composition must be known using the isochron or any equivalent approach before calculation of its date.

The main processes that fractionate elements including parent and daughter elements, and hence can be dated are: (1) Partial melting of a solid; (2) Fractional crystallization of a melt; (3) Metamorphism; (4) Metasomatism; (5) Metal-silicate separation; and (6) Condensation and evapouration. Processes 1 through 5 are well known geological agents. Processes 5 and 6, driven by volatility differences and preferences of elements between Fe-metal and silicate phases, are more relevant in the meteorite context. Other processes in the protoplanetary realm, like accretion of grains into larger bodies and collisions, cause only mixing, and hence cannot be dated directly.

Measurement of a decay interval ending in the present requires direct measurement of the residual parent isotope concentration, indirect determination of its initial concentration, and the use of the basic radioactive decay equation. A decay interval ending in the past will require indirect measurement of both the initial and final concentrations of the parent isotope. Isotopes with long half-lives ($> 10^9 y$) in the freshly isolated solar nebula can be used to measure long decay intervals ending at the present, whereas, isotopes with half-lives ($< 10^8 y$) can be used to measure short decay intervals only at the very beginning of the solar system. As these isotopes have completely decayed by now, they are called extinct isotopes. Their former presence in the Solar System material can be inferred only from their daughter products in the samples, as explained in the last chapter.

Before we consider the various parent-daughter systems and meteorites used in dating the stages in the history of meteorites, we note that the timing of the last event, Event 8 is directly and exactly known, even including the time and day of a meteorite fall. Except as a documentary record, this information may not appear to have any scientific significance. However, Wetherill (1968) was intrigued that afternoon falls of ordinary chondrites far outnumbered the morning falls. This asymmetry in fall time is a consequence of their orbital distributions. His calculations of orbital distribution did not indicate any such asymmetry for objects ejected from the moon, and this ruled out the then-fashionable idea that the moon was the source of ordinary chondrites. The study of lunar samples recovered a year later confirmed this theoretical prediction.

9.2 STAGE 1: FORMATION INTERVALS FROM EXTINCT ISOTOPES

This stage represents what was referred to in the previous chapter as the formation interval. Measurement of this interval requires that a short-lived nuclide survives long after its synthesis in the Interstellar Medium (ISM), is incorporated in the early solids, and leaves its footprint in the form of its daughter isotope. Of the dozen or so short-lived radionuclides, ^{129}I decaying to ^{129}Xe with a half-life of 17 My was considered the most promising candidate in this context. This is because meteorites are likely to incorporate a chemically active element like iodine, but deplete, if not exclude, a chemically inert gas like xenon. Thus, the subsequent and total decay of ^{129}I would leave a distinct ^{129}Xe anomaly or excess. After a few unsuccessful attempts by earlier workers to detect excess ^{129}Xe, John Reynolds of the University of California at Berkeley discovered in 1960 a spectacular excess of ^{129}Xe in the Xe mass spectrum of gases extracted from the H4 chondrite, Richardton, as shown in Figure 9.2 (Reynolds, 1960).

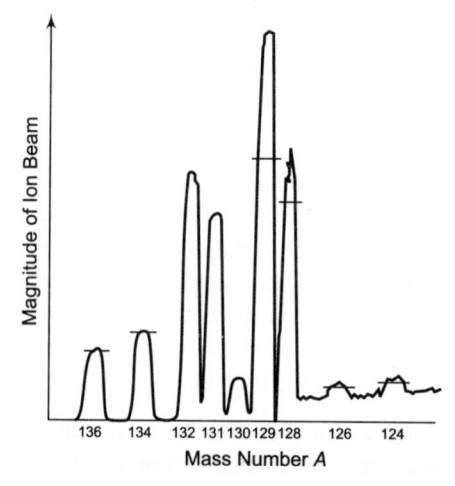

Figure 9.2 The mass spectrum of xenon extracted from the Richardton meteorite (Reynolds, 1960). The horizontal lines show the comparison spectrum of terrestrial Xe. The ^{124}Xe, ^{126}Xe, and ^{128}Xe isotopes have been measured at 10 times the sensitivity of the remaining isotopes

The ratio $(^{129}I/^{127}I)_c$ of residual ^{129}I to the stable ^{127}I at time t_c of meteorite formation and the ratio $(^{129}I/^{127}I)_i$ at the time t_i of isolation from nucleosynthesis will be related by the basic radioactive decay equation as:

$$\left(\frac{^{129}I}{^{127}I}\right)_c = \left(\frac{^{129}Xe^*}{^{127}I}\right)_c = \left(\frac{^{129}I}{^{127}I}\right)_i \exp(-\lambda_{129}\Delta) \tag{9.1}$$

where λ_{129} is the decay constant of ^{129}I, ^{129}Xe* is the excess over the background resulting from the total decay of residual ^{129}I in the meteorite, and Δ the formation interval. Δ can be calculated from the measured $(^{129}I/^{127}I)_c$ if $(^{129}I/^{127}I)_i$ is known or assumed. Jeffery and Reynolds (1961) used an elegant technique to measure the residual $(^{129}Xe^*/^{127}I)_c$ ratio as a single Xe isotope ratio by converting a known fraction of the stable ^{127}I into stable ^{128}Xe by neutron irradiation, and, thereby, measuring the ^{129}Xe/^{128}Xe ratio in different temperature fractions. The typical value of this ratio is

~$1*10^{-4}$. Theoretical estimates of the initial iodine ratio vary from unity to ~$2*10^{-3}$ (Equation 8.7) depending on the assumed time dependence of galactic nucleosynthesis. Corresponding Δ values vary between 200 and 80 My.

The difference of Δ values between any two meteorites (1 and 2, say) can be calculated independent of initial Iodine isotope ratio as:

$$\left(\frac{^{129}Xe^*}{^{127}I}\right)_1 = \left(\frac{^{129}I}{^{127}I}\right)_i \exp(-\lambda_{129}\Delta_1) \qquad (9.2)$$

$$\left(\frac{^{129}Xe^*}{^{127}I}\right)_2 = \left(\frac{^{129}I}{^{127}I}\right)_i \exp(-\lambda_{129}\Delta_2) \qquad (9.3)$$

$$(\Delta_1 - \Delta_2) = \left(\frac{1}{\lambda_{129}}\right) \ln \left[\frac{\left(\frac{^{129}Xe^*}{^{127}I}\right)_2}{\left(\frac{^{129}Xe^*}{^{127}I}\right)_1}\right] \qquad (9.4)$$

The maximum difference in Δ values between primitive meteorites, metamorphosed ordinary chondrites, enstatite achondrites, and even silicate inclusions of iron meteorites was surprisingly less than ~15 my, implying a very 'sharp isochronism' of diverse meteorite groups. The I-Xe studies now focus mainly on this relative quantitative I-Xe chronology of meteorites. The L chondrite, Bjurbole, was earlier used as an arbitrary reference standard for relative I-Xe chronologies. Bjurbole has later been replaced by the aubrite, Shallowater (achondrite) as the reference standard (Brazzle et al., 1999).

The discovery of extinct or fossil ^{129}I prompted a search for the fossil transuranic isotope ^{244}Pu, as it has an even longer half-life (82 vs 17 My), and decays by spontaneous fission into Xe isotopes $^{131-136}Xe$ (other than ^{129}Xe). It was positively identified when the abundance anomalies in ^{132}Xe, ^{134}Xe, and ^{136}Xe found in meteorites matched the mass spectrum of xenon from the spontaneous fission of artificially produced ^{244}Pu. This fission should also produce charged particle fission tracks (Chapter 5) in excess of those expected from uranium. Excess fission tracks were indeed found, but they did not exactly tally with the Xe data. The Pu-Xe formation intervals have been measured like I-Xe ages, but lack of a stable isotope of Pu (analogous to the stable ^{127}I) precluded tracking ^{244}Pu variations that were only due to its decay. Δ values by both systems are, however, broadly consistent.

If the formation interval was indeed as long as 100 My, as indicted by I-Xe or Pu-Xe systematics, the former presence of radionuclides with much shorter half-lives, like ^{26}Al (0.7 My), ^{53}Mn (3.7 My), or ^{60}Fe (1.5 My), would be undetectable in meteorites for three reasons: (1) Their initial abundance relative to their respective stable isotope in the freshly isolated solar nebula would have been lower than $(^{129}I/^{127}I)_i$, unless there was a spike in their production shortly before isolation; (2) They would have become almost, if not completely, extinct before the end of such a long formation interval; and (3) Background concentration of the solid daughter elements [magnesium (Mg), chromium (Cr), and nickel (Ni), respectively] would not be as favourably low as Xenon to reveal any feeble radiogenic daughter isotope (^{26}Mg, ^{53}Cr, and ^{60}Ni, respectively). However, these isotopic pairs have been pursued intensively with positive results. The discovery of fossil ^{26}Al with a much smaller half-life is particularly important for its possible role as an intense heat source in the early solar system. The approach to this discovery begins with the equation for extinct nuclides, derived in Chapter 5 (Equation 5.12):

$$\left(\frac{d}{d'}\right) = \left(\frac{d}{d'}\right)_c + \left(\frac{p_t}{d'}\right)$$

$$= \left(\frac{d}{d'}\right)_c + \left(\frac{p}{p'}\right)_c \left(\frac{p'}{d'}\right) \tag{9.5}$$

Note that the suffix t in Equation 5.12 has been replaced by c to refer specifically to condensation.

For a ^{26}Al-^{26}Mg system, this equation becomes:

$$\left(\frac{^{26}Mg}{^{24}Mg}\right) = \left(\frac{^{26}Mg}{^{24}Mg}\right)_c + \left(\frac{^{26}Al}{^{27}Al}\right)_c \left(\frac{^{27}Al}{^{24}Mg}\right). \tag{9.6}$$

According to this equation, phases within a meteorite with variable (Al/Mg) ratios at the present time will be linearly correlated with their (^{26}Mg/^{24}Mg) ratios, the slope being the residual (^{26}Al/^{27}Al)$_c$ ratio at their formation. G J Wasserburg and his coworkers at the California Institute of Technology (Lee et al., 1976) demonstrated a remarkable correlation of various inclusions in the CV3 meteorite, Allende, as shown in Figure 9.3. The slope of this line is 5.1 ± 0.6 × 10^{-5}. This ratio in the freshly isolated solar nebula is estimated to be between 10^{-3} and 10^{-4}. A residual ratio of 5 × 10^{-5} will then imply that nucleosynthesis occurred no more than a few million years before the formation of Ca- and Al-rich inclusions (CAIs). The former presence of ^{26}Al in meteorites suggests that other extinct nuclides with similar half-lives like ^{53}Mn (3.5 My), ^{60}Fe(1.5 My), ^{107}Pd (9.0 My) and ^{182}Hf (9 My) must also have been present. They have indeed been found (White, 2013). But, the importance of ^{26}Al arises as a heat source very early in the Solar System. If a ^{26}Al/^{27}Al ratio of ~10^{-5} was representative of the bulk of the Solar System, then that would be the dominant heat source in all planetary bodies that formed early (Wasserburg, 1987). It would readily melt and metamorphose bodies only a few kilometres in size. That ^{26}Al could be the primary heat source very early in the Solar System was first proposed by Harold Urey to account for the magmatic activity in planetesimals manifested by differentiated meteorites.

^{26}Al reduces the nebular period to about 10^6 years from 10^8 years inferred from the I-Xe systematics. One possibility is that I-Xe and Al-Mg represent two different nucleosynthetic events. The scenario became blurred when the former presence of ^{41}K with a half life of only 1,50,000 years was also detected also in refractory inclusions of carbonaceous chondrites like Efremovka (Srinivasan et al., 1994). Even more baffling was the detection of the former presence of ^{10}Be in these inclusions, as this nuclide is not produced, but, in fact, destroyed, in galactic nucleosynthesis. So, alternative mechanisms and astrophysical sites are being explored now. For example, short-lived nuclides like ^{10}Be, ^{26}Al, and ^{41}K could be produced by the so called T-Tauri or X-ray star phase of the early sun.

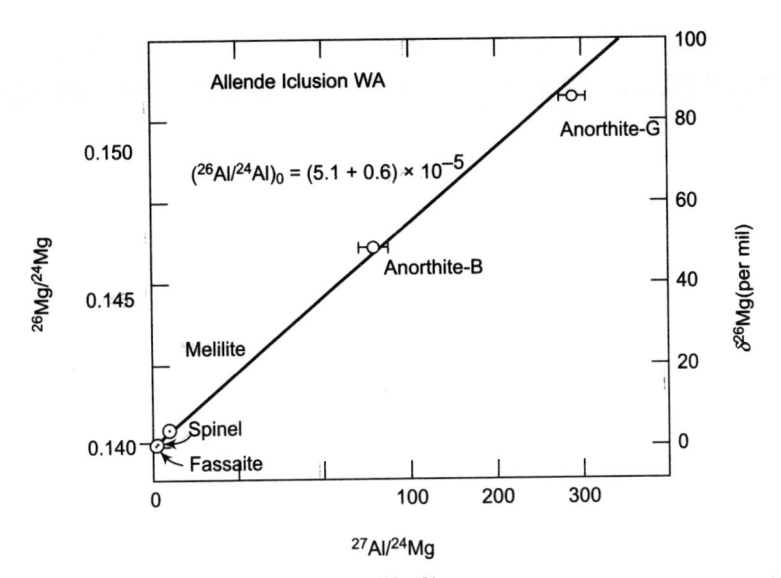

Figure 9.3 The correlation between $^{26}Mg/^{24}Mg$ vs $^{27}Al/^{24}Mg$ for minerals from an inclusion of the Allende meteorite (Lee at al., 1977). The linear correlation implies the *in situ* decay of the short-lived ^{26}Al nuclide.

9.3 STAGES 2 THROUGH 4: FORMATION AGES OF METEORITES

Decay of a long-lived nuclide is used to quantitatively date a variety of events, or processes in the long history of meteorites on the basis of the fundamental age equation in its isochron representation. The isochron ages will refer to processes that equilibrate daughter isotopes on different scales of sampling (Chapter 5). An internal isochron based on sampling on a mineral scale within a meteorite will give the time of the latest isotopic equilibration on a mineral scale, such as metamorphism, or a larger scale, such as magmatic melting, to form differentiated meteorites like achondrites and irons. A whole-meteorite isochron based on sampling individual meteorites will give the time of isotopic equilibration on a much wider scale, presumably in the solar nebula before condensation.

Four long-lived decay pairs, K-Ar, Rb-Sr, Sm-Nd, and U-Pb have commonly been used to date meteorites on the basis of the fundamental age equation (Equation 5.11):

$$\left(\frac{d}{d'}\right) = \left(\frac{d}{d'}\right)_t + \left(\frac{p}{d'}\right)[\exp(\lambda t) - 1] \qquad (9.7)$$

The relevant ratios to be measured for each method are given in Table 9.1.

Of these methods, the K-Ca method is not practical, because of negligible radiogenic addition relative to the initial ratio. The K-Ar method involves a chemically inert gas, and the ages are more like gas retention periods. As these ages are sensitive to even mild thermal disturbances, they are treated separately from those based on Rb-Sr, Sm-Nd, and U-Pb methods. Each of the three methods requires three independent measurements (d/d', p, and d'). In contrast, coupling the two U-Pb decays by dividing their age equations, as explained in Chapter 5, requires the measurement of only daughter isotope ratios (Equation 9.8). Also, the decay constants of the two

Table 9.1 Parent-daughter Ratios in Rb-Sr, Sm-Nd, and U-Pb Systems

Method	(p/d′)	(d/d′)	Decay constant	(d/d′)$_p$
K-Ca	$\dfrac{^{40}K}{^{44}Ca}$	$\dfrac{^{40}Ca}{^{44}Ca}$	$4.96 \times 10^{-10}\,y^{-1}$	–
K-Ar	$\dfrac{^{40}K}{^{36}Ar}$	$\dfrac{^{40}Ar}{^{36}Ar}$	$0.58 \times 10^{-10}\,y^{-1}$	–
Rb-Sr	$\dfrac{^{87}Rb}{^{86}Sr}$	$\dfrac{^{87}Sr}{^{86}Sr}$	$1.42 \times 10^{-11}\,y^{-1}$	0.69899
Sm-Nd	$\dfrac{^{147}Sm}{^{144}Nd}$	$\dfrac{^{143}Nd}{^{144}Nd}$	$6.54 \times 10^{-12}\,y^{-1}$	0.50661
U-Pb	$\dfrac{^{235}U}{^{204}Pb}$	$\dfrac{^{207}Pb}{^{204}Pb}$	$9.85 \times 10^{-10}\,y^{-1}$	10.294
	$\dfrac{^{238}U}{^{204}Pb}$	$\dfrac{^{206}Pb}{^{204}Pb}$	$1.55 \times 10^{-10}\,y^{-1}$	9.307

Uranium isotopes are much larger than those of ^{87}Rb and ^{147}Sm. The combined intrinsic and extrinsic precision of Pb-Pb isochron ages is, therefore, much better than that of Rb-Sr and Sm-Nd ages.

$$\frac{\left[\left(\dfrac{^{207}Pb}{^{204}Pb}\right) - \left(\dfrac{^{207}Pb}{^{204}Pb}\right)_t\right]}{\left[\left(\dfrac{^{206}Pb}{^{204}Pb}\right) - \left(\dfrac{^{206}Pb}{^{204}Pb}\right)_t\right]} = \left(\dfrac{^{235}U}{^{238}U}\right)\dfrac{[(e^{\lambda_5 t} - 1)]}{[(e^{\lambda_8 t} - 1)]} \qquad (9.8)$$

where λ_5 and λ_8 are the decay constants of ^{235}U and ^{238}U, respectively. The numbers in the last column of Table 9.1 are the lowest daughter isotope ratios determined directly or indirectly from meteorites, and are believed to be the ratios of the respective elements in the isotopically homogenized solar nebula. They are referred to as primordial ratios, and hence designated as (d/d′)$_p$. Pb and Sr ratios were directly measured in meteorites or phases so depleted in their (p/d′) ratio that the measured ratio is virtually the original ratio (the first term on the right-hand side of the age equation) with negligible radiogenic addition (the second term). The Sr isotope ratio of 0.69899 was measured in Rb-poor and Sr-rich achondrites (eucrites) by Wasserburg and his coworkers (Papanastassiou and Wasserburg, 1969), and is well known now as the Basaltic Achondrite Best Initial (BABI). The Pb ratios were first measured by Patterson in 1955 in U-poor and Pb-rich troilite (iron sulfide) phase in the iron meteorite, Canyon Diablo. The Pb ratios given in the table are from Tatsumoto et al. (1973). As Sm and Nd are present in comparable amounts in both meteorites and their constituent minerals, the initial Nd ratio of 0.50662 was measured from the y-intercept of well defined internal Sm-Nd isochrons of achondrites.

9.4 Rb-Sr, Sm-Nd AND U-Pb AGES OF METEORITES

A large number of meteorites have been dated by one or more of the three methods: Rb-Sr, Sm-Nd, and U-Pb. A comprehensive list of the dates reported before 1990 is given by Dalrymple (1991) with original references. Internal isochron results for about 25 meteorites taken from Dalrymple (1991) are given in the Table 9.2. The table includes primitive, metamorphosed (petrological type 4 to 6), and differentiated meteorites like achondrites and irons. Obvious general remarks on the data in the table are: (1) Analytical errors in Rb-Sr and Sm-Nd ages are large,~ 30 Ma; (2) Pb-Pb ages are more precise, about 5 Ma for the reasons mentioned above; (3) No precise internal isochrons for carbonaceous chondrites; (4) Rb-Sr ages of iron meteorites (Colemara and Weekeroo Station) are not from the bulk iron, but small silicate inclusions in them; and (5) Only one chondrite, St.Severin, and one achondrite, Juvinas have been dated by all three methods, though not in the same laboratory. Broad conclusions emerging from the table are: (1) No differences beyond experimental errors between Rb-Sr, Sm-Nd, and Pb-Pb ages; (2) Internal isochron ages of primitive, metamorphosed, and differentiated meteorites are indistinguishable within the analytical errors. Thus, initial condensation, later metamorphism on, and magmatic melting in parent bodies were all complete within ~100 Ma between 4.55 and 4.45 Ga; and (3) Pb-Pb ages are somewhat older than Rb-Sr and Sm-Nd ages, but this could be due to larger uncertainties in the decay constants of ^{87}Rb and ^{147}Sm.

Table 9.2 Rb-Sr, Sm-Nd and U-Pb Internal Isochron Ages of Meteorites

Meteorite	Type, Grade	Rb-Sr (Ga)	Sm-Nd (Ga)	Pb-Pb (Ga)
Allende	CV3			4.530 ± 0.004
Guarenas	H6	4.46 ± 0.08		
Tieschitz	H3	4.52 ± 0.05		
Barwell	L5			4.554 ± 0.005
Jelica	LL6	4.42 ± 0.04		
St. Severin	LL6	4.51 ± 0.05	4.55 ± 0.33	4.543 ± 0.019
Krahenberg	LL5	4.60 ± 0.03		
Soka Banja	LL4	4.45 ± 0.02		
Parnalee	LL3	4.53 ± 0.04		
St. Saveur	E5	4.46 ± 0.05		
Indarch	E4	4.39 ± 0.04		
Juvinas	Eucrite	4.40 ± 0.07	4.56 ± 0.08	4.540 ± 0.001
Momoa	Eucrite		4.46 ± 0.03	
Sierra de Mage	Eucrite		4.41 ± 0.02	
Stannern	Eucrite	4.48 ± 0.07		
Y-75011	Eucrite	4.46 ± 0.06	4.55 ± 0.04	
Angra dos Reis (ADR)	Angrite		4.56 ± 0.04	4.544 ± 0.002
Colomera	Iron	4.51 ± 0.04		
Weekeroo station	Iron	4.39 ± 0.07		

Table 9.3 Rb-Sr (Ga) and Pb-Pb Whole-Meteorite Isochron Ages

Meteorite	Rb-Sr (Ga)	Pb-Pb (Ga)
Carbonaceous chondrites, C	4.37 ± 0.34	
Ordianry chondrites, L	4.44 ± 0.12	4.551 ± 0.007
Ordinry chondrites, LL	4.486 ± 0.020	
Ordianry chondrites, H	4.512 ± 0.039	
Enstatite chondrites, E	4.508 ± 0.037	4.577 ± 0.004.
C + H + L + E		4.505 ± 0.016
H + L		4.562 ± 0.10
H + L + Eucrites		4.484 ± 0.030
Eucrites	4.30 ± 0.25	4.52 ± 0.02
Eucrites + howardites		4.54 ± 0.02

As pointed out earlier, one possible way to distinguish the primary from secondary ages is to use whole-meteorite isochrons. Table 9.3 lists a few whole-rock Rb-Sr and Pb-Pb isochron ages for intergroup and intragroup meteorites taken from Dalrymple (1991). Whole-rock Rb-Sr age for carbonaceous chondrites is very imprecise, possibly due to breach of isochron conditions. Allegre and his coworkers have used high precision Rb-Sr measurements of 17 H chondrites, 13 LL chondries, and 8 E chondrites to report ages and initial Sr ratios for each group as follows (Minister et al., 1982):

17 H chondrites	4.518 ± 0.039 Ga, Sr ratio, 0.69876 ± 0.00040
13 LL chondrites	4.486 ± 0.020 Ga, Sr ratio, 0.69887 ± 0.00012
8 E chondrites	4.508 ± 0.037 Ga, Sr ratio, 0.69880 ± 0.00044

An isochron combining all 38 data sets gives an even more precise age of 4.498 ± 0.015 Ga and initial Sr ratio of 0.69885 ± 0.00010, as shown in Figure 9.4. The tightness with which all 38 specimens fit the same isochron is very remarkable, and indicates that ordinary chondrites and enstatite chondrites formed within 10–20 Ma of each other from the solar nebula. The failure of L chondrites to tightly fit this combined isochron is believed to be due to the disturbance of their Rb-Sr systematics by shock heating (Section 9.7). Whole-rock Pb-Pb ages are slightly older, but the common Rb-Sr age could be 4.55 Ga within the uncertainty in the value of the ^{87}Rb decay constant. Very limited spread in Sm/Nd ratios between chondrites precludes their whole-rock isochrons.

One of the intergroup ages in Table 9.2 is of historical significance: the Pb-Pb age measured by Patterson based on an isochron (Figure 9.5) defined by one H chondrite, Forest City, one L chondrite, Modoc, one achondrite, Nuevo Laredo, and troilite (FeS) from the Fe meteorite, Canyon Diablo (Patterson, 1956). The original age of 4.55 ± 0.07 Ga is now corrected to 4.484 ± 0.030 Ga (Table 9.2) using more precise values for U decay constants, and troilite composition. This is the first ever age for meteorites, and implicitly also for the Solar System formation.

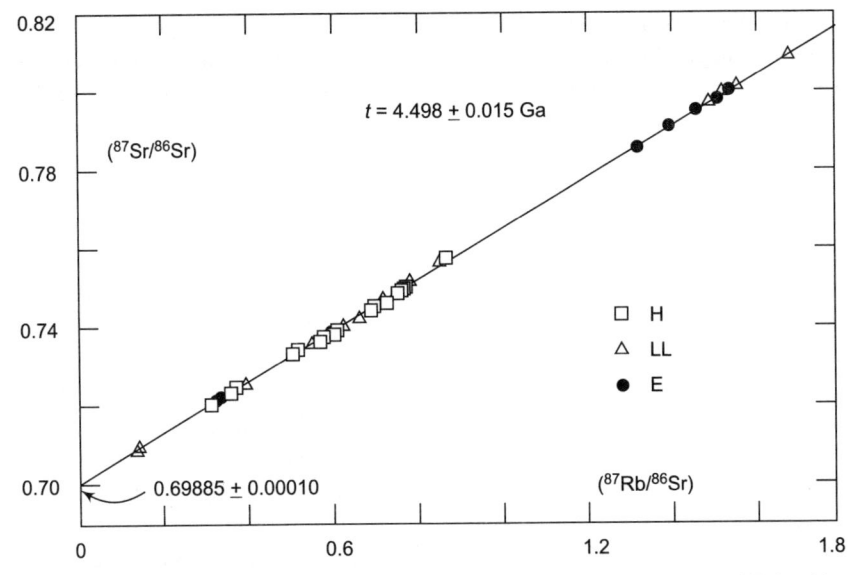

Figure 9.4 Rubidium-Strontium whole-rock isochron for 38 undisturbed H, L, LL, and E chondrites (Minster et al., 1982).

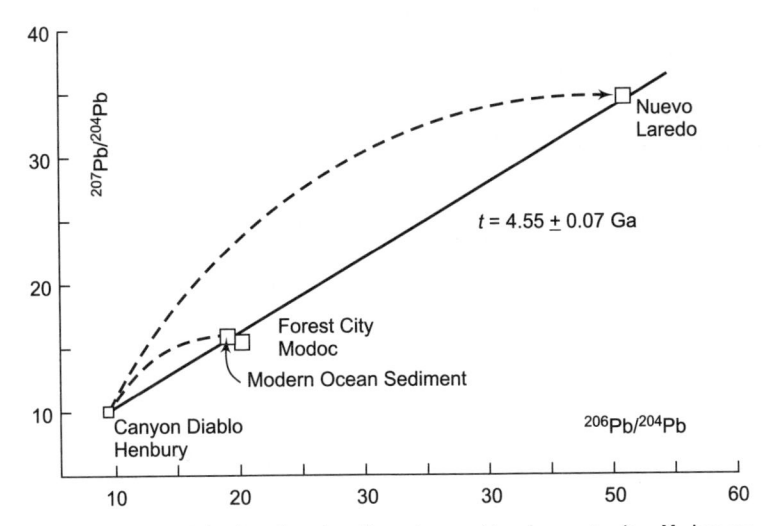

Figure 9.5 Pb-Pb meteoritic isochron based on three stone and two iron meteorites. Modern sea sediments also fall on the isochron (Patterson, 1956).

9.5 Very High Precision Model Ages

It is clear from tables 9.2 and 9.3 that precision of Rb-Sr and Sm-Nd internal isochron ages is insufficient to resolve early Solar System events within 20 My. Even the more precise Pb-Pb isochron ages are barely adequate. We noted earlier that initial Sr and Pb ratios can be measured to very high precision (last column of Table 9.1) by analysing meteorites or their phases with negligible or very low (p/d'). In other words, the second term on the right-hand side of the age equation is zero or negligible relative to the first term, so that the present-day ratio is virtually the initial ratio. This measurement is free of any assumption. Using the same strategy, age can be measured to very high precision by analysing a meteorite or its phases with very high (p/d') ratios, and assuming, necessarily, a value for the initial ratio (the first term). This so-called model age is, in effect, based on a two-point isochron, the points being the assumed initial ratio, and that of the meteorite. Improvement in precision of Rb-Sr and Sm-Nd model ages is, however, limited by the absence of any meteorite phase with a high enough (p/d'), and the need to measure the parent and daughter isotope concentrations with a precision much less than that of an isotope ratio. But by coupling the age equations for the two U-Pb decay pairs into a single equation (Equation 9.8), a model age calculation requires only isotopic analysis of the sample lead. Dalrymple (1991) has compiled a list of Pb-Pb model ages with a precision of ~0.1 per cent for many meteorites, even from whole rock. Selecting phases with much higher U/Pb ratio, Allegre and his coworkers in their Paris laboratory have achieved analytical precisions of less than one million years in 4,550 million years (Gopel et al., 1994). Such incredible precision is equivalent to resolving just three days at the birth or conception of a 45.5 year old person (Wasserburg, 1987). At such high precisions, interlaboratory differences can blur small, but significant age differences. Thus, results from the same lab, as given in Table 9.4, will be less equivocal.

Allegre and his coworkers measured high precision Pb-Pb model ages of high U/Pb phases: (1) Refractory inclusions from the Allende meteorite, believed to be the earliest condensate from the solar nebula; (2) Phosphate fractions produced by post-accretional thermal processing of ordinary chondrites; and (3) Mineral and lithic fragments in basaltic achondrites. One would intuitively expect samples from (2) and (3) to substantially postdate samples from 1. Their results are given in Table 9.4 and summarized in Figure 9.6. The main conclusions are (Gopel et al., 1994; Allegre, 2001):

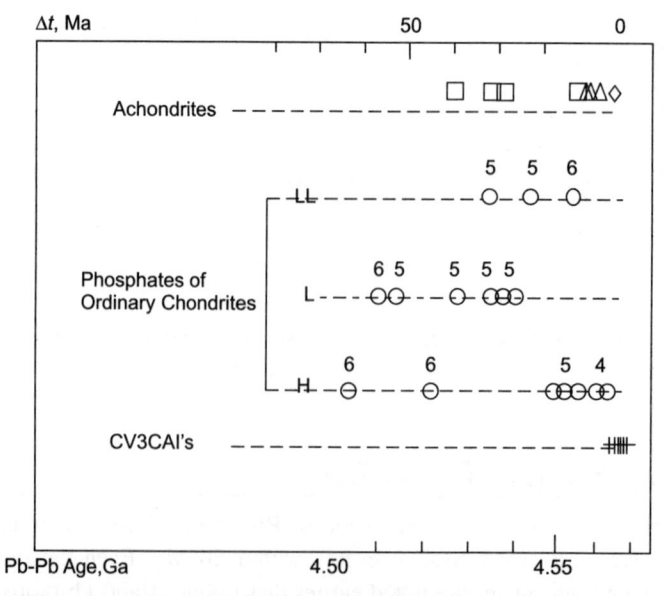

Figure 9.6 High precision Pb-Pb ages of Allende CAIs, Ordinary chondrites, and Achondrites (Allegre, 2001).

1. Allende inclusions with an age of 4567 ± 1 Ma are the oldest yet known Solar System material as inferred from the condensation sequence.
2. Ordinary chondrites formed within ~ 3 My after the Allende inclusions, but some of them underwent metamorphic recrystallization lasting up to ~ 60 My.
3. Melting, segregation of basaltic magma, and its eruption to the surface of parent bodies were all completed within 8 My after the Allende inclusions.

Recent results of high precision Pb-Pb and U-Pb ages are listed in Table 9.5. They confirm the previous results, and add that chondrules formed within a few million years after the Allende inclusions, and that Re-Os ages of iron meteorites are similarly old.

It is incredible that condensation from the solar nebula, accretion of condensates into planetesimals (asteroids), and their igneous differentiation were all completed within 10 My in the early Solar System.

Table 9.4 Pb-Pb Model Ages of Meteorite Phases with Very High U-Pb Ratios

Meteorite	Type	Whole Rock/Phase	Age (Ma)
Alllende	CV3	CAI	4568
Guarena	H6	phosphate	4505
Kernouve	H6	"	4521
Allegan	H5	"	4550
Richardton	H5	"	4551
Nadiabondi	H5	"	4556
Ste.Marguerite	H4	"	4563
Forest Vale	H4	"	4561
Marion	L6	"	4511
Barwell	L5-6	"	4538
Ausson	L5	"	4527
Knyahinya	L5	"	4540
Homestaed	L5	"	4514
St.Severin	LL6	"	4554
Guidder	LL5	"	4535
Tuxtuac	LL5	"	4544
Angra dos Reis	Angrite		4558
Lewis Cliff 86010	Angrite		4558
Ibitira	Eucrite		4556 ± 6
Juvinas	Eucrite		4539 ± 3
Bereba	Eucrite		4536
Nuovo Laredo	Eucrite		4529 ± 5
Bouvante	Eucrite		4510 ± 4

Table 9.5   More Recent Results of Pb-Pb Model Ages

Meteorite	Type	Phase	Pb-Pb age (Ma)	Reference
NWA 2364	CV3	CAI	4568	Bouvier et al., 2010
Allende	CV3	CAI	4568	Bouvier et al., 2007
Efremovka	CV3	CAI	4567	Amelin et al., 2009
Allende	CV3	chondrule	4566	Bouvier and Wadhwa, 2010
Gujba	CB	chondrule	4565	Bouvier and Wadhwa, 2010
Ste. Marguerite	H4	phosphate	4563	Bouvier et al., 2007
D'Oribigny	angrite	phosphate	4564	Amerlin, 2008
ADR	angrite		4558	Amelin, 2008
Lewis Cliff 86040			4559	Ameelin, 2008
Asuka881394	eucrite		4566	Wadhwa et al., 2009
NWA 2976	eucrite		4563	Bouvier et al., 2011

9.6   Meteorite Ages Much Younger than 4.5 Ga

A Rb-Sr age of 3.73 Ga supported by a similar K-Ar age of silicate inclusions from the iron meteorite, Kodaikanal is the first indication that not all iron meteorites formed as early as 4.5 Ga, for example, Colomera and Weekeroo Station in Table 9.2. An identical Ar-Ar age of 3.79 Ga of silicate inclusions from another iron meteorite, Neschevao, suggests that both iron meteorites originated from metal-silicate fractionation on the same parent body ~ 700 Ma after Solar System formation. A real surprise was Rb-Sr and Sm-Nd ages as young as 1.3 and 0.180 Ga from the rare subgroup (SNC for Shergotty, Nakhla and Chassigny) of achondrites. The SNC achondrites are Ca-poor unlike Ca-rich eucrites, and resemble terrestrial diabase, dunite, and pyroxenite. Their crystallization from magma generated so late in the Solar System history requires parent bodies with protracted energy sources and, hence, larger than asteroids. That Mars, rather than asteroids, was the most probable source of SNC achondrites came from the similarity of the noble gas structure of the Martian atmosphere analysed by the Viking Landers in 1975, with that of the meteorites. Dynamical considerations support ejection of relatively unshocked and intact pieces of rock with escape velocities by impacts on planetary surfaces. If impact debris can reach the earth from asteroids and Mars, those from the nearby Moon are also to be expected, and indeed are found among the meteorites.

9.7   Stage 5: Gas Retention Ages and Post-Formational Cooling and Heating Histories

Radiometric dating methods based on the decay of long-lived nuclides to gaseous daughter elements are K-Ar, U-He, Th-He, and U-Xe (Faure and Mensing, 2005). Of these, the K-Ar method has widely been applied to meteorites. Being chemically inert, Ar is retained or immobilized in meteoritic minerals only at much lower temperatures than the chemically active Sr, Nd, and Pb. Thus, gas retention ages by the K-Ar method could be distinctly younger than Rb-Sr, Sm-Nd, and U-Pb (Pb-Pb) ages for slowly cooling bodies. For the same reason K-Ar ages can be far more easily disturbed and even reset, than Rb-Sr and Sm-Nd ages by heating events subsequent to primary cooling. The

K-Ar ages have, therefore, been used chiefly to date either slow cooling or secondary heating events. Partial argon loss due to relatively mild heating episodes will result in spurious ages that cannot be easily detected. Fortunately, a technical variant of the K-Ar method—^{40}Ar-^{39}Ar method—includes self checks on the validity, or otherwise, of calculated ages, as explained below. The age equation for K-Ar decay is (Section 5.3):

$$\left(\frac{^{40}Ar}{^{36}Ar}\right) = \left(\frac{^{40}Ar}{^{36}Ar}\right)_t + \left[\left(\frac{\lambda_e}{(\lambda_e + \lambda_\beta)}\right)\right]\left(\frac{^{40}K}{^{36}Ar}\right)[\exp(\lambda t) - 1] \quad (9.9)$$

where λ_e and λ_β are decay constants for the branching electron-capture and beta (β) decays, and $\lambda_e/(\lambda_e + \lambda_\beta) = C_1$ is the fraction of ^{40}K atoms decaying to ^{40}Ar. The first term is usually not the initial Ar ratio, but recent contamination from atmospheric Ar, and, hence, can be subtracted from the left-hand side to give the radiogenic ^{40}Ar in absolute amount as:

$$^{40}Ar = C_1{}^{40}K \cdot [\exp(\lambda t) - 1]$$

Or,

$$t = \left(\frac{1}{\lambda}\right)\ln\left[1 + \left(\frac{1}{C_1}\right)\left(\frac{^{40}Ar}{^{40}K}\right)\right] \quad (9.10)$$

The Ar-Ar technique eliminates separate analysis of K by converting a fraction C_2 of ^{39}K into ^{39}Ar by the fast neutron reaction ^{39}K$(n, p)^{39}$Ar in a nuclear reactor. Then, ^{39}Ar $= C_2.^{39}$K $= C_2.C_3.^{40}$K, where C_3 is the (^{39}K/^{40}K) ratio. C_2 will depend on the flux, reaction cross section, and energy range of neutrons and irradiation time.

$$^{40}Ar = J.^{39}Ar [\exp(\lambda t) - 1]$$

Or,

$$t = \left(\frac{1}{\lambda}\right)\ln\left[1 + J\left(\frac{^{40}Ar}{^{39}Ar}\right)\right] \quad (9.11)$$

J is determined simply by co-irradiating a standard (flux monitor) of known age t_m with the sample as:

$$J = \frac{\left[\exp(\lambda t_m) - 1\right]}{\left(\frac{^{40}Ar}{^{39}Ar}\right)_m} \quad (9.12)$$

Corrections for atmospheric Ar and other interfering Ar isotopes produced by irradiation are necessary, but straightforward. Calculation of t from Equation 9.9 requires separate measurements of K and Ar in the bulk sample, whereas, Equation 9.11 gives it from just an Ar isotope ratio. This means that the sample can be heated to progressively higher temperatures, and an age can be calculated from the Ar ratio of each gas fraction. A plot of age vs gas fraction, called an age spectrum, is very informative, as each gas fraction represents the Ar/K ratio in different mineralogical or lattice site in the bulk sample. If all sites in a sample retained Ar quantitatively since its formation, all gas fractions will give the same age and define a 'plateau' as shown in Figure 9.7a (Dalrymple, 1991). The plateau age must be the true age of the sample. Figure 9.7b shows a plot of ^{40}Ar/^{36}Ar vs ^{39}Ar/^{36}Ar isotope data for each fraction, called an ^{40}Ar-/^{39}Ar isochron or correlation diagram

(Dalrymple, 1991). The data will fall on a straight line whose slope is equal to the radiogenic ^{40}Ar/^{39}Ar ratio and whose y-intercept is the ^{40}Ar/^{36}Ar ratio of the trapped argon. The isochron plot does not require the assumption that the nonradiogenic, or trapped argon is atmospheric.

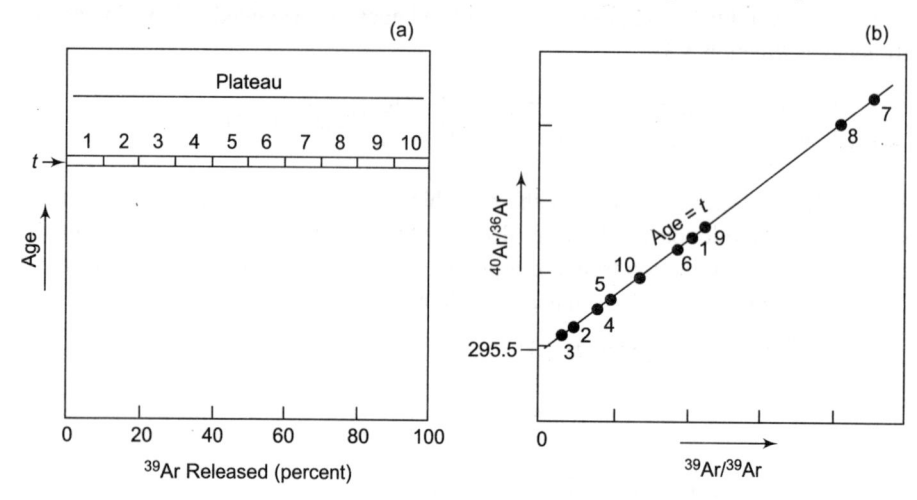

Figure 9.7 Hypothetical ^{40}Ar-^{39}Ar age spectrum. Each of the gas increments (1 through 10) gives the same age t (a), and ^{40}Ar/^{36}Ar isochron plot with the y-intercept giving the trapped argon (b) (Dalrymple, 1991).

For a sample that is heated at some later time $t1'$ after its original formation, some sites may have lost some argon, and, hence, the gas fraction from these sites will give younger Ar-Ar ages. The retentive sites sampled at higher temperatures may still define a plateau corresponding to the true age, as shown in Figure 9.8a, and an isochron, as shown in Figure 9.8b (Dalrymple, 1991). Disturbed samples may and do give patterns more complicated than the one shown in Figure 9.8. Such disturbed age spectra give insights into the thermal history of meteorites. The Ar-Ar technique

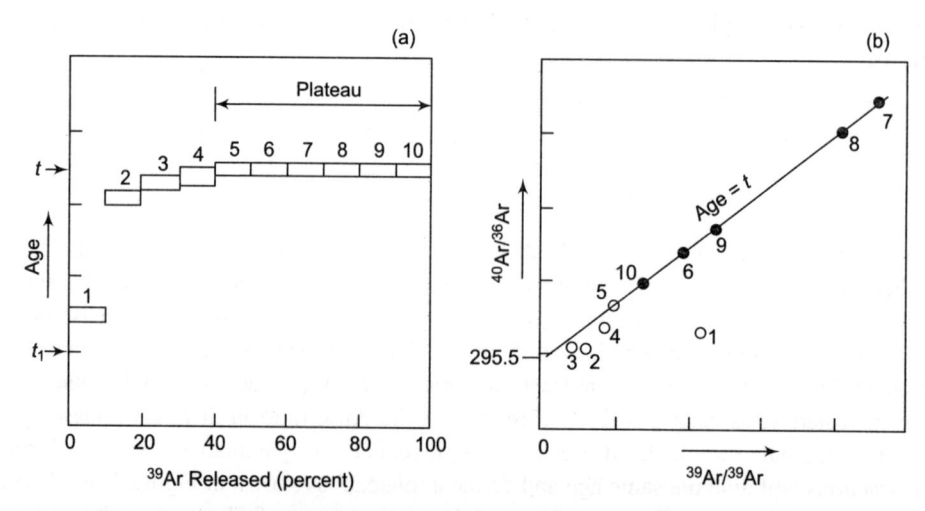

Figure 9.8 Age spectrum of for a hypothetical sample that lost argon from less retentive sites due to later (t') heating (a), and isochron for this sample (b) (Dalrymple, 1991).

provides valid ages even if the sample as a whole was not perfectly closed. It is now the preferred method for K-Ar dating of both terrestrial and extraterrestrial rocks (McDougall and Harrison, 1988).

Conventional K-Ar ages of unshocked meteorites, determined before the development of the Ar-Ar method, are generally similar to ages given by other methods. Many other meteorites from all groups gave distinctly younger K-Ar ages that were interpreted as due to their later heating from energetic collisions (Anders, 1964). Intensely shocked L chondrites gave U-He and K-Ar gas-retention ages between 1.3 and 0.4 Ga. A tighter cluster of K-Ar ages of the black L chondrites at ~500 Ma has been interpreted as the time of a single major collisional breakup of the parent body of L chondrites (Heymann, 1967). The ability of the Ar-Ar technique to distinguish between strongly and poorly retentive Ar sites in meteorites was evident from the Ar-Ar ages of many meteorites (Dalrymple, 1991). Unshocked meteorites (chondrtes, achondrites and irons) gave ages ranging from 4.53 ± 0.03 Ga (indistinguishable from Rb-Sr, Sm-Nd, and Pb-Pb ages) to 4.38 ± 0.03 Ga. Gopal et al. (1994) interpreted this time span of 150 My as representing the cooling time of parent bodies of different size and/or an early period of brecciation and shock reheating during the accretion of planetesimals. Strongly shocked meteorites gave age spectra or Ar release patterns that are complex in details, but commonly have a low temperature age plateau marking the time of the shock event. Recent Ar-Ar dating of one of the shocked L chondrites gave the age of the shock event as 470 ± 6 Ma. This precise age may have remained untested had it not been for the chance discovery of as many as 40 highly altered chondrites, identified as of L group, in a thin limestone layer precisely dated at 467.3 ± 1.6 Ma (early Ordovician) at Kinnekule, Sweden (Schmitz et. al, 2001). This close similarity of ages is very convincing evidence that the L chondrite parent body was disrupted by an impact close to 470 Ma. Fragments from that breakup were transferred through gravitational resonance into earth-crossing orbits. These fragments must have rained down on the earth at a rate of '25 times the current meteorite flux rate' for several million years (Schmitz et. al, 2001).

9.8 STAGE 6: DURATION OF METEORITES AS SMALL INDEPENDENT OBJECTS IN SPACE

We have so far considered the history of individual meteorites when they were an integral part of much larger bodies—their parent bodies. From the various ages of meteorites, we inferred, based on short-lived and long-lived nuclides, that parent bodies went through accretion of small grains, recrystallization after accretion in the solid state (metamorphism) in surficial regions, and melting (magmatism) in the deeper parts, followed by sporadic collisional reheating.

We will now consider the history of a meteorite as an independent small body from the time it is ejected from its parent body into interplanetary space to its eventual capture by the earth. All objects in space are subject to irradiation by energetic charged particles in the cosmic rays. As explained in chapters 2 and 5, these charged particles, mostly protons and alpha (α) particles, originate in the sun (solar cosmic rays) and in the interstellar medium (galactic cosmic rays). Both solar and galactic cosmic rays show a power-law distribution of energies, from ~10 to a few hundred MeV in the former and from 10^2 to 10^4 in the latter. Low energy particles in the solar wind

(~1 KeV/nucleon) are only implanted in the first few atomic layers of solids. Energetic heavy particles penetrate more deeply and disrupt the crystal lattice leaving behind 'tracks' that can be viewed under high magnification in a Transmission Electron Microscope and under low magnification when etched in suitable reagents (Chapter 5). Partilces with much higher energies (of several MeVs) induce nuclear reactions by breaking the target nuclei into two or more smaller nuclides (spallation), which eject many secondary neutrons. As these secondary neutrons slow down to low velocities (thermalized), they are efficiently absorbed by the target nuclei to produce a variety of, both stable and unstable isotopes called cosmic-ray-produced or 'cosmogenic' nuclides (Chapter 2). Table 9.6 lists few typical stable and unstable cosmogenic nuclides. Production of cosmogenic isotopes increases with depth to a peak between 0.5 and 1 metre below surface. Thus, only the first 1 metre or so of a meteorite will accumulate cosmogenic stable isotopes, while unstable isotopes will reach a steady state.

Table 9.6 Cosmogenic Isotopes in Meteorites (McSween and Huss, 2010)

Nuclide	Half-life (Y)	Produced from
^{3}H	12.26	O, Mg, Si, Fe
^{3}He	stable	O, Mg, Si, Fe
^{10}Be	1.5×10^6	O, Mg, Si, Fe
^{14}C	5730	O, Mg, Si, Fe
^{21}Ne	stable	Mg, Al, Si, Fe
^{22}Na	2.6	Mg, Al, Si, Fe
^{26}Al	7.17×10^5	Si, Al, Fe
^{38}Ar	stable	Fe, Ca, K
^{36}Cl	3.01×10^5	Fe, Ca, K
^{81}Kr	2.29×10^5	Rb, Sr, Y, Zr

Using laboratory measurement of production rates of spallation and neutron capture nuclides as a function of depth and measured ^{3}He, ^{21}Ne, and ^{36}Ar in meteorites, it is straight forward to calculate the cosmic ray exposure age of a meteorite, but for the loss of its preatmospheric size and shape by ablation during atmospheric entry. Thus, it is necessary to reconstruct the preatmospheric size and shape to locate the position of the recovered meteorite within it in order to use the appropriate production rates for age calculation. Because of this and many other complications, like contributions of previous irradiations, cosmic ray exposure ages are not very accurate. Cosmic ray exposure ages have been measured for various classes of meteorites (McSween and Huss, 2010). We will consider only findings (Wieler and Graf, 2001). Iron meteorites have the widest range of ages from 0 to 2.4 Ga with a notable cluster between 0.2 and 1.2 Ga. Chondrite ages are typically much less than 100 My, and are distributed differently for different groups. The difference between irons and chondrtis is mainly due to their physical strength. In contrast to the large scatter in exposure ages of chondrites, Martian meteorites (about 40 falls) indicate three discrete impact ejection events. Exposure ages of lunar meteorites indicate impact events in the lunar highlands and mare regions.

9.9 STAGE 7: TERRESTRIAL RESIDENCE TIME OF METEORITE FINDS

This stage refers to the time from the unnoticed fall of a meteorite somewhere on the earth to its recovery by humans by sheer chance or concerted and intelligent search. The criterion for recognition of meteorites fallen in the past is their strangeness to their immediate environment, be it farm lands, barren deserts, or icy terrains.

The so called 'terrestrial age' of a meteorite is determined from the cosmogenic isotopes produced by exposure to cosmic rays in free space. Exposure to cosmic rays ceases when the meteorite falls on the earth and is shielded by the earth's atmosphere. During the cosmic ray exposure, abundances of stable cosmogenic isotopes will increase with time, whereas, those of the unstable or radioactive nuclides like ^{14}C, ^{26}Al, ^{10}Be, and even ^{36}Cl will reach a steady state between production and decay. The steady abundance of unstable cosmogenic nuclides will decay exponentially according to their respective decay constants in a fallen meteorite. By comparing their residual abundance in a meteorite find now with the expected steady state value, its terrestrial age can be calculated. A ^{14}C decay will be useful for terrestrial ages up to 30,000 years, while ^{26}Al, ^{10}Be, and ^{36}Cl are useful for much longer (a few million years) terrestrial ages.

Terrestrial ages of meteorites found in desert sands range up to about 40,000 years, whereas, terrestrial ages for meteorites found in Antarctica and other ice-covered terrain range up to 5,00,000 years (McSween and Huss, 2010). Jull (2001) has summarized the data on terrestrial ages. Discovery of fossilized and highly altered L chondrites with a terrestrial age of 470 million years in a limestone quarry in Sweden was mentioned earlier. Terrestrial ages provide information on meteorite survival relative to weathering in different climatic regions on the earth.

10 Chemical Evolution of the Earth

> In astronomy we see things not as they are but as they were. In geology, we see things not as they were but as they are now.
>
> **Leon Long**

> The temporal sequence is converted into a simultaneous co-existence, the side-by-side existence of things into a state of mutual interpenetration ... a living continuum in which time and space are integrated.
>
> **Tibetan monk, Lama Govinda's religious experience as cited by P Davies**

10.1 Composition of Terrestrial Planets and Chondritic Meteorites

The close similarity in the abundances of elements heavier than oxygen in chondritic meteorites and the solar atmosphere (Figure 1.5) suggests a chondritic composition as a reasonable assumption for the composition of the inner rocky planets—Mercury, Venus, Earth, and Mars. However, even simple considerations show that there may be important differences in chemical composition between meteorites and the inner planets. For example, Table 10.1 gives the densities of the four terrestrial planets and chondritic meteorites (Brown and Mussett, 1993). The table shows that the terrestrial planets have densities, which when corrected for their different internal pressure, vary significantly from that of chondritic meteorites. In fact, Mars with an uncompressed density of ~ 3,700 kg m^{-3} is the only planet which lies in the chondritic range of 3,400–3,900 kg m^{-3}. Earth, Venus, and Mercury are denser than chondrites, and the Earth's moon is less. The only sufficiently abundant element to influence such strong density difference is iron, because its atomic mass 56 is much higher than that of the next three most abundant elements, O(16), Mg(24), and Si(28).

Table 10.1 Densities of Terrestrial Planets and Chondritic Meteorites

Object	Density (kg m^{-3})	Estimated Uncompressed Density (kg m^{-3})
Mercury	5,435	5,300
Venus	5,245	4000
Earth	5,514	4000
Moon	3,344	3,300
Mars	3,934	3,700
Chondrites	3,400–3,900	3,400–3,900

More subtle is the evidence that volatile elements have been systematically depleted in terrestrial planets. Figure 10.1 shows a plot of the ratio of volatile elements, K and Rb to refractory (less volatile) elements U and Sr, respectively, in meteorites, inner planets, and the moon (Halliday and Porcelli, 2001). This shows the volatile depleted nature of the moon relative to the Earth, and the Earth and Mars relative to meteorites. The most depleted ratio in angrites is understandable as they are differentiated meteorites (Chapter 9).

The systematic decrease in iron metal/silicate ratios and systematic variations in ratios of volatile elements to refractory trace elements were probably caused by decreasing condensation pressures and temperatures, respectively, in the solar nebula away from the sun (Brown and Mussett, 1993). This also means that the systematic variations must have occurred during or after condensation, but before accretion. The exception to the systematic decrease in densities is the moon. This suggests that the moon did not form along with terrestrial planets directly from the solar nebula.

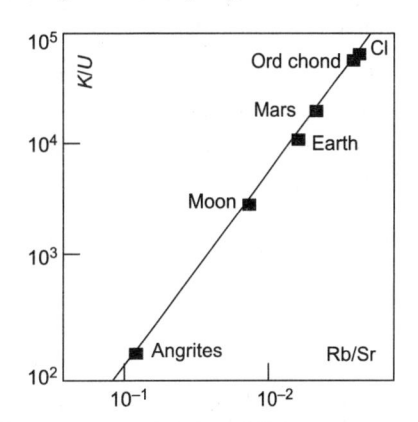

Figure 10.1 The ratio of volatile elements (K, Rb) to refractory elements (U, Sr) in planetary and Solar System objects (Halliday and Porcelli, 2001).

10.2 ENERGETIC PROCESSES DURING THE FINAL STAGES OF EARTH ACCRETION

According to the standard model of planetary accretion (Wetherill, 1990), planets accreted in three distinct stages—coagulation of microscopic mineral grains in the solar nebular disc into planetesimals (like the parent bodies of meteorites), then the gravitational agglomeration of planetesimals into a few hundred planetary embryos, that is, the two initial stages on a relatively short time scale (10^4–10^5 years), and finally collisional amalgamation of embryos on a longer time scale of 10^7–10^8 years into the few planets as observed today. In other words, giant impacts between embryos are essential to the final stages of planetary assembly. This implies, in the specific context of the Earth, that:

1. The Earth experienced one or more giant impacts during its accretion and reached nearly its present size within 10^8 years after the formation of the earliest condensate at $4,567 \pm 1$ Ma (Chapter 9);

2. Impacts would have almost inevitably led to planet-wide melting and attendant formation of a sustained and deep magma (molten rock) ocean;
3. The moon, in fact could have originated from an impact between the Earth and a large planetary embryo;
4. Separation of any excess iron from the magma ocean, and its gravitational sinking into an inner iron core, overlain by a silicate slag or residue in the embryonic earth; and
5. Degassing of volatile gases and components during core formation to form an early terrestrial atmosphere and ocean.

Although no direct evidence for the one-time presence of a magma ocean is now available, circumstantial evidence is strong. For instance, there are indications (Chapter 11) that the core, early atmosphere, and the primitive ocean formed during or soon after the Earth's accretion, linking all the three layers to a magma ocean.

The silicate blanket overlying the iron core is termed the Bulk Silicate Earth (BSE) or Primitive Mantle (PM), and is an important reference composition for the study of mantle evolution. For this reason, major and trace element compositions of the BSE have been estimated on the basis of several assumptions (Rollinson, 2007). The chondrite model for the BSE is based on the composition of chondrites, adjusted for the depletion of volatile elements and for the segregation of iron and other siderophile elements into the core (Zindler and Hart, 1986).

The subsequent evolution of the BSE has been driven by two sources of energy, the external solar energy and the internal heat of the Earth. The solar energy intercepted by the Earth has remained more or less steady at the current value of 10^{17} watts. The bulk of this energy is radiated back into space with only a minute fraction used to drive the circulation of the present-day atmosphere and ocean and, through that, the physical and chemical weathering and transport processes, resulting in erosion and sedimentation. Solar energy would have acted in a similar way on any atmosphere or ocean as soon as they formed on the early Earth system. The present-day Earth's internal heat generation [made up of residual gravitational energy from formation, heat generated from radioactive decay of U and K (Chapter 2), and heat (latent) released from the solidification of the inner core] is estimated to be equal to the measured heat escaping from the interior of the earth at $\sim 4 \times 10^{13}$ watts. This internal heat energy must have been much higher in the past because of higher residual gravitational energy and higher abundances of radioactive elements. This is the fundamental energy source for the dynamics of the mantle (Chapter 2).

The stratification of the BSE into many layers with distinct chemical compositions resulted from preferences of chemical elements to different types of materials and states of matter combined with melting processes in the mantle. It is, therefore, instructive to consider the chemical affinities of elements and the role of melting processes in the transfer of elements from one host to another.

10.3 ELEMENT SEGREGATION: SOME GEOCHEMICAL RULES

V. M. Goldschmidt originally grouped the major elements in the Earth into four major categories on the basis of his findings on meteorites. Those elements with an affinity for silicates or oxygen are lithophile, those with an affinity for sulphur are chalcophile, those with an affinity for metallic iron are siderophile, and those with an affinity for the gaseous atmosphere are atmophile. The first three chemical affinities are now explained by their differing electronegativities (Pauling, 1988). Electronegativity (E) is defined as a dimensionless number between 0 and 4 representing a quantitative measure of the ability of an atom to attract electrons and, hence, become a negatively charged anion. Electronegativity increases from left to right and decreases from top to bottom of the periodic table (Figure 1.2). The most electronegative cations are at the bottom left and the most electropositive anions at the top right. The lithophile, chalcophile, and siderophile elements are grouped according to their E values as (Brown and Mussett, 1993):

Lithophile $E < 1.6$ (K, Na, Ca, Mg, Mn, Al, Ti, and complex-forming ions of B, Si and P)
Chalcophile $1.6 < E < 2.0$ (Zn, Fe, Co, Ni, Pb and Cu)
Siderophile $2.0 < E < 2.4$ (As, Pt, Ir, Au, etc.)

Elements combine with the same and other elements by means of three types of bond—ionic, covalent, and metallic. Lithophile elements form ionic bonds with oxygen of very different E value (3.5). Chalcophile elements form covalent bonds with sulphur with $E = 2.5$. Covalent bonds are characterized by sharing of electrons between the bonded atoms. In metallic bonding, as between transition metals, electrons are free to move among the bonded atoms.

Iron with an E value of 1.8 should normally be chalcophile. The reason for its siderophilic affinity in the Earth's environment is due to the relative abundances of the major elements in the Earth, the best estimate in atom percent being: O (30.7), Mg (14.2), Al (1.4), Si (15.5), S (3.8), Ca (1.5), and Fe (32.90. These seven elements account for nearly 97 per cent of the Earth's mass. As discussed by Brown and Mussett (1993), all the lithophile cations Si^{+4}, Mg^{+2}, Al^{+3}, and Ca^{+2} (total atom per cent, ~32) will combine with oxygen to form various silicates like olivine and pyroxene. But because oxygen is far more abundant, some of it will remain unused. The normally chalcophile Fe will combine with S to form FeS. But the remaining Fe will combine with residual O and, hence, become lithophile. The iron left after combining with oxygen and sulphur will remain as free metal to become the host to the siderophile elements in the Earth. Thus, if the early Earth had acquired sufficient energy to melt it substantially (as envisaged by the standard model of planetary accretion), major elements in it would have been sorted out into lithophile, chalcophile, and siderophile layers. Even an embryonic earth could have differentiated due to melting by the energy of accretion and short-lived radioactivities.

10.4 Segregation of Major and Trace Elements During Melting or Igneous Processes

Igneous rocks are formed from molten material known as magma, which usually consists of a solution of the Earth's most abundant elements, oxygen and silicon, with smaller amounts of aluminum, calcium, magnesium, iron, sodium and potassium (Philpotts, 1990). Robert Bunsen was the first to recognize, in 1851, that magmas were actually 'solutions' no different from those of salts in water, except that they were hotter. Most magmas are silicate melts, but rarer ones contain little or no silica, and are composed essentially of calcium carbonate, sulfide, or iron oxide. The major elements of common magmas combine, on cooling, to form the so-called rock-forming minerals—quartz, feldspars, felspathoids, pyroxenes, olivine, and garnet and, in the presence of water, amphiboles and mica. In addition, common accessory minerals, such as iron-titanium oxides, apatite, zircon, and sulfides, form from minor magmatic constituents [Fe^{+3}, Ti, P, Zr, and S, respectively] that do not readily enter the structures of the major rock-forming minerals.

The history of an igneous magma begins with its formation at some depth in the earth. The composition of the magma is determined by the chemical and mineralogical composition of the rock in the source region and by the process of melting. When sufficient melt has been formed and coalesced, buoyancy causes it to rise. Further chemical modifications may occur during this period of transport. The magma may rise to the surface and extrude as lava. Most magma, however, solidifies beneath the Earth's surface, where slow cooling and crystallization allow for further modification of the initial composition. During this period, a rock develops its characteristic appearance, which is referred to as texture.

Melting of multi-mineral planetary material fractionates, or segregates, both major and trace elements, but by different processes, as explained briefly below. Mixtures of minerals do not melt at a single temperature, but over a range of temperatures, because of the different melting points of individual minerals. This range is bounded by the inception of melting, called the solidus, and the completion of melting at the liquidus. Below the solidus, the mixture is solid. Between the solidus and liquidus, the system is only partially melted, consisting of a mixture of unmelted minerals and liquid, that is, partial melt. The composition of the partial melt is different from the original bulk composition. Thus, separation of the melt (magma) from the solid residue due to buoyancy forces will result in the fractionation of major elements. The solidus temperature is always lower than the melting points of any of the constituent minerals. The melt produced at the solidus temperature of a multiphase system is saturated in all components making up the phase assemblage and is called the eutectic melt. The chemical composition of this eutectic melt is determined by the phases present, but not their relative abundance. That means, rock containing (say) 90 per cent olivine, 5 per cent clinopyroxene, and 5 per cent spinel will produce the same eutectic melt as one with 33.3 per cent of each phase. The properties of melting mixtures have been extensively covered in standard text books on igneous petrology (Ragland, 1989; Philpotts, 1990). The solidus temperature of a multicomponent system generally increases with confining pressure, and decreases substantially with its water and volatile content.

As opposed to just about half a dozen major elements (concentration > 1 per cent) there are at least 70 elements in the BSE and meteorites in very low concentrations—from parts per million (ppm) to parts per billion (ppb) levels, and, hence, are called trace elements. Yet, many of these trace elements have an importance in igneous and other processes like metamorphism and sedimentation that far outweigh their abundance. Their very scarcity makes them very valuable in understanding processes like generation of magmas at depth, and their transfer to or close to the surface of the Earth and subsequent crystallization. Unlike major elements, trace elements rarely form minerals of their own, and, therefore, seek 'sites' in the available solid and liquid phases. The distribution between minerals and any fluid (sea water, fresh water, or magma) is extremely sensitive to parameters such as the composition of the fluids, oxidation state, temperature, and to a lesser extent, pressure, in addition to the chemistry of the minerals. The behavior of trace elements in igneous processes is commonly discussed in terms of their distribution between the different minerals, the melt, and any volatile-rich phase that may be present. More formally, the distribution of a trace element between coexisting mineral and liquid phases is defined by its partition coefficient, D_i, as:

$$D_i = \frac{C_m^i}{C_l^i} \tag{10.1}$$

where, C_m^i and C_l^i are the concentrations in weight units of element i in the mineral and melt (liquid) phases, respectively. The value of D_i depends on temperature, pressure and melt, and mineral compositions. The value of the D reflects the ease with which an element can substitute in a particular mineral. The ease of substitution of a trace element in a mineral depends on the similarity of its ionic radius and charge to those of the major elements in the crystal. When the solid phase consists of more than one mineral, an effective partition coefficient, called the bulk distribution coefficient, $\bar{D}_i$, can be calculated as:

$$\bar{D}_i = \Sigma W_i \cdot D_i \tag{10.2}$$

where, W_i is the weight fraction of each mineral in the solid (the so called mode of the solid) and D_i is the mineral/melt distribution coefficient of the element for each mineral. $\bar{D}_i$ may be viewed as the distribution coefficient of a single fictitious phase that has the same distribution coefficient as the weighted average of the mineral assemblage.

Elements with $D < 1$ are termed incompatible; they will be preferentially concentrated in the liquid phase during melting and crystallization. In contrast, those with $D > 1$ are called compatible and these will be preferentially retained or extracted in the residual or crystallizing phases respectively. Which elements are compatible in a rock depends on its mineralogical composition.

10.5 Graphical Representation of Inter-Element Variations in Compatibility

Geochemists use a variety of trace element diagrams to display multi-element data. Typically, a multi-element diagram will display a range of elements equally spaced along the x-axis of the diagram according to a particular property, like incompatibility (Rollinson, 2007). The y-axis depicts the abundances of those elements commonly on a logarithmic scale to accommodate a large range of abundances. The abundances are normalized to the concentrations in an appropriate reference material. These reference materials are mostly chondritic meteorites for the Rare Earth Elements (REEs) and an estimate of the composition of the Earth's primitive mantle (or BSE) for many other trace elements (Rollinson, 2007). The REEs are the lanthanides, elements 57 (La) to 71 (Lu) in the periodic table. They are of particular interest because they are geochemically very similar. All carry +3 charge and they show smoothly decreasing ionic radii from La to Lu. They are expected, therefore, to behave as a coherent group and show smooth and systematic variation in their compatibility through the series (Figure 10.2). The REE plots have been widely used to understand magmatic processes in igneous and other types of rocks. On a primitive-mantle-normalized trace element diagram elements are not ordered according to their atomic number, unlike in the case of REEs. Instead, they are arranged in order of decreasing incompatibility during mantle melting. In other words, those elements with the greatest preference for the melt phase during mantle melting are plotted to the left of the graph and those with a lesser affinity to the right, as shown in Figure 10.3. Here the purpose is to see how the sample understudy differs from the primitive mantle from which it was presumably derived.

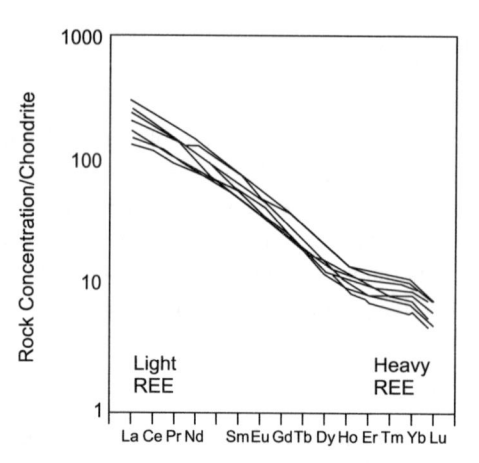

Figure 10.2 The graph showing the enrichment of the rare earth elements in rocks relative to chondritic meteorites. The light rare earth elements are more enriched than the heavy rare earth elements (Rollinson, 2007).

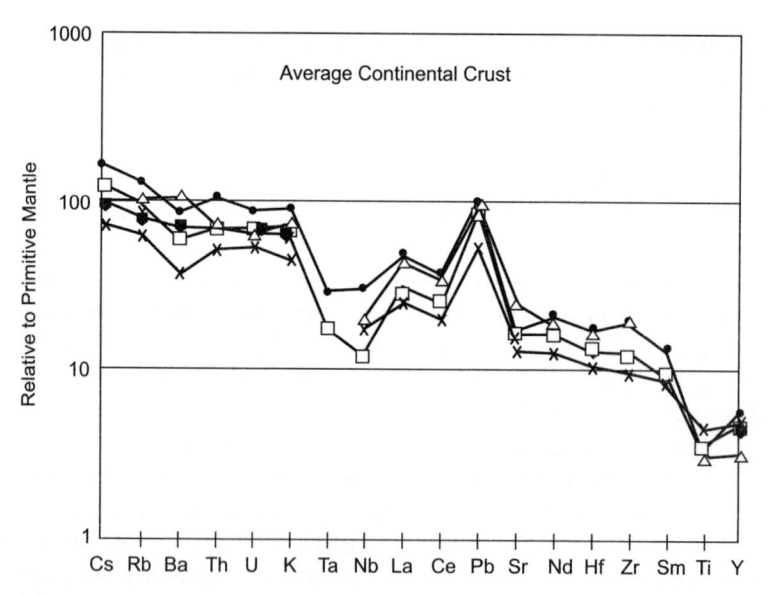

Figure 10.3 A primitive-mantle-normalized multi-element plot for selected trace elements in average continental crust. The elements are arranged in order of decreasing incompatibility (Rollinson, 2007).

10.6 MELTING AND CRYSTALLIZATION MODELS

Having defined how elements are partitioned between minerals and melts, their behavior during melting and crystallization can be represented by simple mathematical expressions. The reader is referred to many text books (Ragland, 1989; White, 2013) for the derivation of these expressions, as we shall simply quote them in the following discussion.

10.6.1 Partial melting

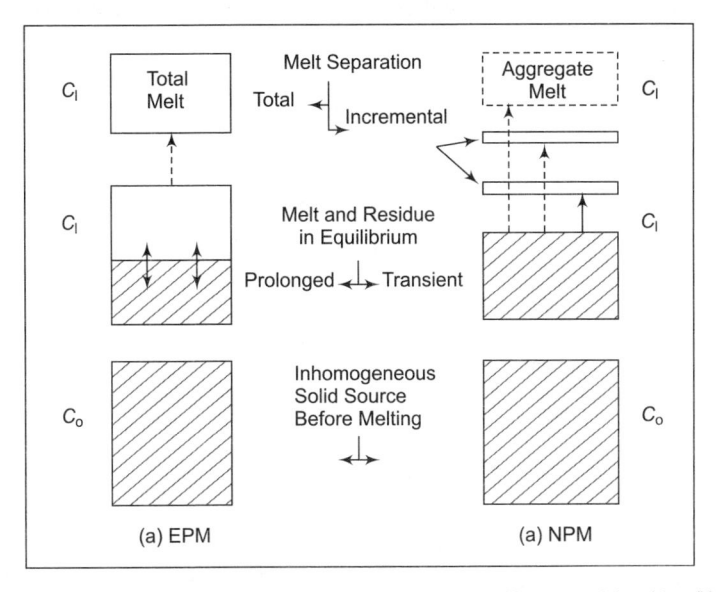

Figure 10.4 Equilibrium partial melting (a), and Non-equilibrium partial melting (b).

The simplest model for partial melting of a complex mineral assemblage is known as the Batch melting or Equilibrium Partial Melting (EPM) model, shown schematically in Figure 10.4. It describes the formation of a partial melt which remains in contact with and in chemical equilibrium with the solid residue (shown by the double arrows) until it is sufficiently buoyant to separate as a single batch of magma. Under these circumstances the concentration of an element in the liquid (C_l) relative to that in the initial unmelted source (C_0) is given by (Wetherill et al., 1981):

$$\left(\frac{C_l}{C_0}\right) = 1/[F + (1-F)\bar{D}_R] \quad (10.3)$$

where, F is the fraction of partial melting (equal to the amount of melt/amount of initial solid) and $\bar{D}_R$ is the bulk partition coefficient in the residual solid. The degree of enrichment or depletion (C_l/C_0) for different values of F is illustrated in Figure 10.5 for different values of $\bar{D}_R$. When $\bar{D}_R$ is small, the expression reduces to ($1/F$) and marks the limit of trace element enrichment for any given degree of batch melting. When F is small, the expression reduces to ($1/\bar{D}_R$) and marks the maximum possible enrichment of an incompatible element and the maximum depletion of a compatible element relative to the original source. Small degrees of melting can cause significant changes in the ratio of two incompatible elements with different $\bar{D}_R$s. The concentration in the melt relative to the source for compatible elements with $D > 1$ approximates to ($1/\bar{D}_R$) throughout much of the melting range, especially for very high values of $\bar{D}_R$, and only increases towards the limit of unity for very high values of F. Very high values of F are unlikely, as the melt would probably be buoyantly unstable after about 30 per cent melting. The concentration of elements (both compatible and incompatible) in the refractory residue is always given by $\bar{D}_R \cdot C_l$ and it follows that, even for very small degrees of melting the residue becomes severely depleted in incompatible elements, whereas, compatible elements remain close to their initial concentrations.

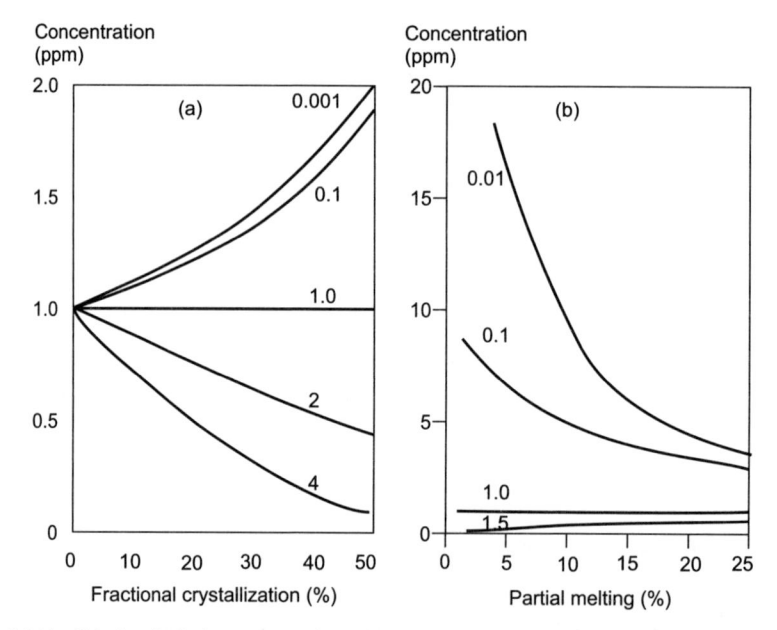

Figure 10.5 Calculated relative concentrations of trace elements in a liquid after different degrees of partial melting and fractional crystallization. The original solid or liquid, respectively, contained 1 ppm of a trace element, and different curves reflect particular distribution coefficients.

Non-equilibrium Partial Melting (NPM), also known as Rayleigh Fractional Melting, refers to melting in which an infinitely small amount of liquid is formed in equilibrium with the residue and then removed from the system, as shown schematically in Figure 10.4. Although this melting may be physically unreasonable as a model for magma production, it serves to illustrate the limiting extremes of changes in the trace element contents of melts and residues where the melt migrates rapidly in comparison with diffusion rates in the solid phases. The concentration of a trace element in the liquid relative to the parent rock for a given melt increment is given by:

$$\left(\frac{C_l}{C_0}\right) = \left(\frac{1}{\bar{D}}\right)(1-F)^{\left(\frac{1}{\bar{D}}-1\right)} \tag{10.4}$$

where, F is the fraction of melt already removed from the source and $\bar{D}$ is the bulk partition coefficient for the original solid phase prior to onset of melting. The changes in concentrations are more extreme than in batch melting. The limit to enrichment of incompatible elements in the melt is still $(1/\bar{D})$, but that of compatible elements tends to infinity as F approaches 1. The melts are still in equilibrium with the residue (if only momentarily) so that $C_r = \bar{D} \cdot C_l$ once more. It follows that incompatible elements are more efficiently purged from the source than in the batch melting. In the range 0–10 per cent melting, the changes in element concentrations relative to the original source are more extreme than in batch melting, although the limiting value $(1/D)$ is the same.

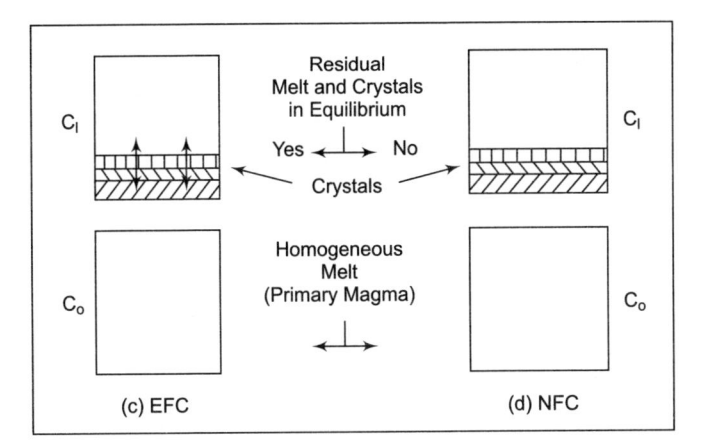

Figure 10.6 Equilibrium fractional crystallization (a), and Non-equilibrium fractional crystallization (b).

A variant of NPM is one in which the increments are pooled (upper most box in dotted outline) and thoroughly mixed before migrating buoyantly upward. This is in practice numerically indistinguishable from batch melting except for the behavior of compatible elements at very high degrees of melting.

10.6.2 Crystallization of a homogeneous melt

Unless the magma is very rapidly quenched into a glassy matrix in its final resting place on or close to the surface of the Earth, it will evolve from its initial composition (both major and trace elements) as crystals nucleate from it. As in partial melting, two extreme types of fractional crystallization are conceivable: Equilibrium Fractional Crystallization (EFC) and Non-equilibrium Fractional Crystallization (NFC), also known as Rayleigh Fractional Crystallization, shown schematically in Figure 10.6. EFC refers to all the crystalline products remaining in the liquid and re-equilibrating continuously (shown by the double arrows) as the liquid slowly crystallizes. This is simply a reversal of batch melting or EPM, and gives the concentration C_l of a trace element in the liquid relative to the parent concentration C_0 as:

$$\left(\frac{C_l}{C_0}\right) = \frac{1}{[f + (1-f)\bar{D}_F]} \tag{10.5}$$

where, f is the fraction of the melt remaining and D_F is the bulk partition coefficient of the crystalline assemblage. The enrichment and depletion of a trace element relative to the original liquid are shown in Figure 10.5. EFC is considered to be an unlikely process in nature, as zoned crystals would probably grow faster than they could re-equilibrate with the changing liquid composition and gravitationally settle as cumulates. NFC corresponds to the more likely scenario, namely the magma being isolated in a magma chamber and the crystals being separated from the liquid as rapidly as they form, precluding any re-equilibration with the changing liquid composition. The concentration of a trace element in the liquid relative to the initial concentration is now given by:

$$\left(\frac{C_l}{C_0}\right) = f^{(\bar{D}_F - 1)} \tag{10.6}$$

NFC is illustrated in Figure 10.5 for different $\bar{D}_F$. It is far less effective than EPM for concentrating an incompatible element relative to another.

The incompatible behavior of an element is not constant, but may vary significantly over the crystallization history of the magma. A particularly spectacular example of such behavior is that of Zr dissolved in a basaltic melt. Over much of the basalt's crystallization history, Zr behaves incompatibly and is concentrated in the melt. With a distribution coefficient of 0.01 for Zr with the major silicate minerals like olivine, pyroxene, and plagioclase crystallizing from the basalt magma, an initially low concentration of ~100 ppm will increase to about 5,000 ppm after 98 per cent of the melt has crystallized. Experiments show that at about this point the mineral zircon begins to crystallize, as the distribution coefficient between zircon ($ZrSiO_4$) and the melt is enormous (Hess, 1989). This may not be by itself remarkable, but for the selective inclusion of U and rejection of Pb in the nucleating zircon mineral. Nucleation of zircon will selectively incorporate U and exclude Pb to give very high ratios of U/Pb. This makes zircon an extremely useful geochronometer.

The EPM and NPM are extreme models of partial melting as are EFC and NFC models of crystal fractionation from a liquid. More realistic models must be intermediate between these extremes, with or without additional complications. For example, it is unlikely that any process of partial melting works so efficiently that all of any melt is purged from the source rock. Some melt probably stays behind and mixes with the next batch.

10.7 Combined Partial Melting and Recrystallization

In the ideal case of formation of a partial melt from an upper mantle source at depth by EPM, its transfer without any contamination en route to the surface, and its eventual chemical differentiation due to NFC, we can combine equations 10.3 and 10.6, and write the concentration of an element in a lava sample as (Wetherill et al., 1981):

$$C_l = \frac{C_0 f^{(\bar{D}_F - 1)}}{[F + (1 - F)D_R]} \qquad (10.7)$$

One way of estimating the five unknown parameters (C, F, f, D_r, D_f) is to use plausible values for these parameters, calculate C_l using the above equation, and see how well it agrees with the observed C_l. A satisfactory agreement for a representative set of trace elements will increase confidence in the model known as forward or direct modelling. But with so many free parameters to choose from, the solution may not be sufficiently unique to allow firm conclusions to be drawn.

A different approach, quite commonly used in geophysics and called inverse modelling, is to make use of the variations in the observed elemental concentrations in a suite of cogenetic rocks to determine unknown parameters, such as the composition and mineralogy of the source, the physical processes of partial melting and fractional crystallization causing the observed pattern of trace element distributions in the rock suite, and the degree of partial melting and crystal fractionation. This approach was pioneered by Allegre and his colleagues in Paris (Allegre, 1987). This approach is like determining the history of basaltic magma from the observed data, whereas, the forward approach is like predicting the output for basaltic magma with an assumed history. The inverse approach is believed to lead to improved uniqueness of the solutions.

10.8 Observational Constraints on the Structure and Composition of the Modern Mantle

10.8.1 The internal structure of the earth

The internal structure of the Earth is revealed primarily by velocities of compressional (*P*) and shear (*S*) waves that pass through the Earth from earthquake centres. Such seismic wave velocities vary with pressure (depth), temperature, mineralogical composition, chemical composition, and the degree of partial melting. Three first-order seismic discontinuities divide the Earth into crust, mantle, and core—the Mohorovicic discontinuity or Moho defining the base of the crust; the core-mantle interface at 2,900 kilometre; and inner core-outer core interface around 5,200 kilometre. These discontinuities reflect changes in composition, phase or both. Smaller, but important velocity changes at 50 to 200 kilometre, 410 kilometre, and 660 kilometre are the basis for further subdivision of the mantle. The major divisions of the Earth (Figure 10.7) can be summarized as follows (Condie, 2005):

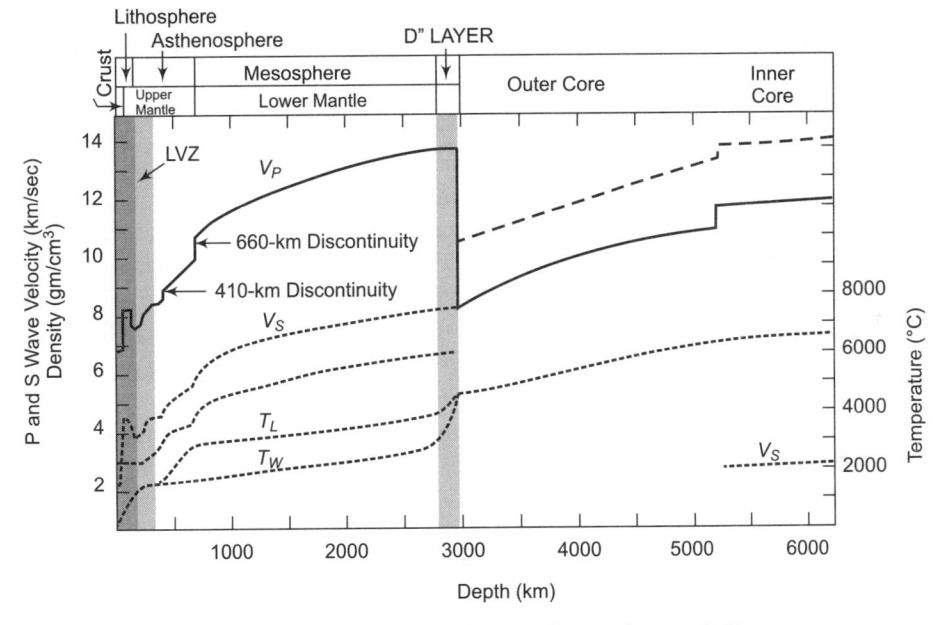

Figure 10.7 The distribution of average *P*-wave, *S*-wave velocities, and calculated average density in the earth (Condie, 2005).

1. The crust consists of the region above the Moho with a thickness of about 3 kilometre at some oceanic ridges and about 70 kilometre in collisional orogens.

2. The lithosphere (50–300 km thick) is the strong outer layer of the Earth, including the crust, which reacts to many stresses as a brittle solid. The asthenosphere, extending from the base of the lithosphere to the 660 kilometre discontinuity, is by comparison a weak layer that slowly deforms by creep and can, therefore, convect. A region of low seismic velocity and of high attenuation of seismic wave energy, the Low Velocity Zone (LVZ), occurs with

varying thickness (50–100 km) at the top of the asthenosphere. Significant lateral variations in density and in seismic wave velocity are common at depths of less than 400 kilometre.

3. The upper mantle extends from the Moho to the 660 kilometre discontinuity and includes the lower part of the lithosphere and the upper part of the asthenosphere. The region from 410 kilometre to the 660 kilometre discontinuity is known as the transition zone. These two discontinuities demarcate a region within the mantle in which the crystal structures of the constituent minerals are transformed into denser phases.

4. The lower mantle extends from the 660 kilometre to the 2,900 kilometre discontinuity at the core-mantle boundary. It is characterized by rather constant increases in velocity and density in response to increasing hydrostatic compression. Between 200 and 250 kilometre above the core-mantle interface, a flattening of velocity and density gradients occurs in a region known as the D" layer. The lower mantle is also referred to as the mesosphere, a region that is strong and relatively passive in terms of deformational processes.

5. The outer core will not transmit shear waves and is, hence, interpreted to be liquid. It extends from 2,900 kilometre to the 5,200 kilometre discontinuity.

6. The inner core, extending from the 5,200 kilometre discontinuity to the centre of the Earth transmits shear waves, although at very low velocities, suggesting that it is a solid near the melting point.

There are only two layers in the Earth with anomalously low seismic-velocity gradients: the lithosphere and the D" layer just above the core. These two layers coincide with steep temperature gradients; hence, they are thermal boundary layers in the Earth. Both layers play an important role in the cooling of the Earth. Most cooling (> 90 per cent) occurs by plate tectonic mechanisms (see later) as plates are subducted deep into the mantle. The D" layer is important in that steep thermal gradients in this layer may generate mantle plumes (next section), many of which rise to the base of the lithosphere, thus bringing heat to the surface (< 10 per cent of the total Earth cooling).

10.8.2 Plate tectonics and mantle convection

Plate tectonics, developed in the early 1970s has profoundly influenced geologic thinking ever since. Before this theory was developed, earthquakes, volcanoes, variations in the magnetic field over the ocean, seafloor bathymetry, the distribution of fossils on different continents, the distribution of ancient glacial deposits, and mountain belts all appeared to be features unrelated to one another. Plate tectonics unified these diverse features as various surface manifestations of a single global process called sea floor spreading (Langmuir and Broecker, 2012).

Plate tectonics is now understood as driven directly or indirectly by mantle convection. The following is a brief description of the coupling between convection and plate tectonics to show how crust and heat are extracted from the mantle, and oceanic crust is cycled back into the mantle (Fowler, 1990). Figure 10.8 is a representation of the mantle-crust coupling.

The surface of the Earth is divided into 10 major lithospheric plates and several smaller ones. Each plate moves as a rigid unit relative to adjoining ones. Where plates move apart due to mantle convection, new material must fill the intervening space. Thus, along the mid-ocean ridges (middle of Figure 10.8), which are divergent plate boundaries, the rising asthenosphere decompresses and partially melts to form basaltic magma that buoyantly rises and solidifies to create new lithosphere. Elsewhere (right and left of the middle), material must be consumed to conserve the surface area of

Chemical Evolution of the Earth

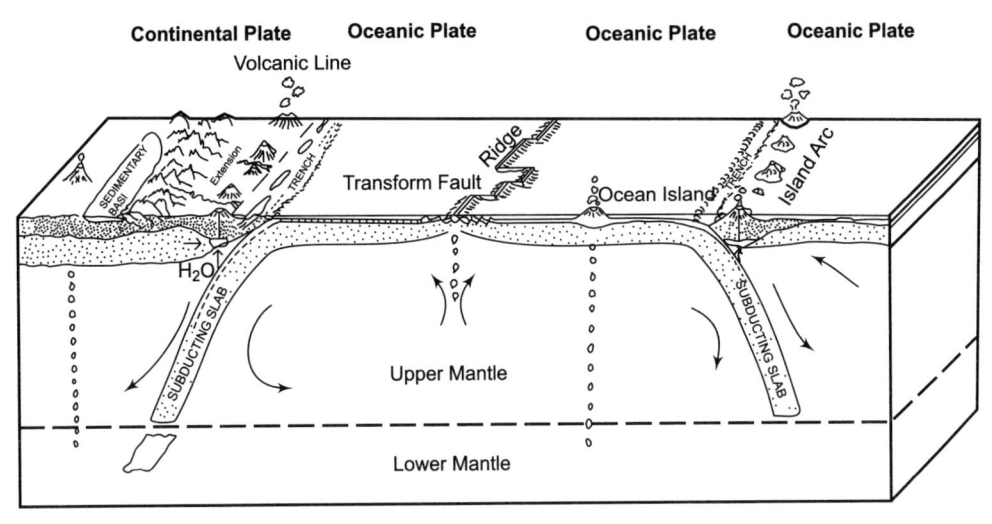

Figure 10.8 Schematic illustrating the formation of new oceanic lithosphere along mid-ocean ridges and its eventual subduction into the mantle. Lithosphere is stippled. Crust is indicated by dense stipple. Ocean island basalts may be derived from the lower mantle (Thompson, 1990; Fowler, 1990).

the Earth. This occurs at convergent plate boundaries where cold lithospheric slabs are subducted into the mantle. Plates may also slip past one another along transform faults, as happens on the San Andreas Fault. As the new hot lithosphere moves away from mid-ocean ridges, it cools due to circulation of cold ocean water and is hydrothermally altered in the process. The lithosphere thickens, becomes denser, and sinks deeper into the asthenosphere, thus causing oceans to deepen away from ridges (trenches such as the Marianas Trench). At subduction zones, the sinking of cool lithospheric plates into the asthenosphere depresses isotherms to considerable depths, and as a result, oceanfloor rocks are converted into low-temperature high-pressure metamorphic rocks. This metamorphism releases water and carbon dioxide (CO_2), which rises into the overlying mantle wedge where (especially water) it serves as flux to fuse and form basaltic and andesitic (higher silica than basalts) magmas. These lighter magmas buoyantly rise to the surface into the base of the continental crust (right in Figure 10.8). Here, dense basaltic magma may intrude or underplate the crust. As it cools, it can melt the overlying rocks to form bodies of granitic magma that rise through the crust to form batholiths and possibly erupt on the surface as rhyolitic volcanoes. Andesitic magma may continue all the way to the Earth's surface to from cone-shaped volcanoes.

Although the plates are made up of both oceanic and continental material, usually only the oceanic part of any plate is created or destroyed. At subduction zones, where continental and oceanic materials meet, it is the oceanic plate which is subducted (and thereby destroyed). The continents are rafts of lighter material which remain on the surface, whereas, the denser oceanic lihthosphere is subducted beneath either oceanic or continental lithosphere. Continents cannot be subducted. Instead, the subduction zone ceases to operate at that place and moves to a more favourable location. Mountains are built above subduction zones as a result of continental collisions. The realization that plates can include both oceanic and continental parts, but that only the oceanic parts are created or destroyed, removed the main objection to the theory of continental drift proposed early in the last century by A Wegener.

Most volcanism and crustal deformation occur along plate boundaries. A few volcanoes, however, are located in plate interiors, as in Hawaii. As ocean islands often form long lines of volcanoes of progressively increasing ages they are believed to be created by an apparently stationary heat source (hot spot) under a moving plate. Morgan proposed that hotspots are created by narrow cylindrical columns of hot mantle plumes rising from the deep mantle. Whether the plumes rise from the base of the upper mantle or the base of the lower mantle (as depicted in Figure. 10.8) remains an unsettled issue.

The above description shows that basaltic magmas are extruded from the mantle at three main settings in which mantle melting occurs. Basalts extruded at mid ocean ridges are called Mid-Ocean-Ridge Basalts (MORB), at island arcs of convergent plates are called Island-Arc Basalt (IAB), and basalts extruded in the interior of lithospheric plates are called Ocean-Island Basalts (OIB). Volcanic rocks equivalent of OIBs extruded on continents are known as Continental-Flood Basalt (CFB). Production of MORB at about 4 km^3 per year far exceeds the production of other two basalts at the present time. A MORB may also have been dominant in the past. Chemical and isotopic analyses of the three groups of volcanic rocks and other basaltic melts on the Earth's surface provide an important 'window' into mantle evolution.

10.8.3 Mineralogical makeup of rocks in the upper mantle

While the oceanic and continental crusts can, in principle, be sampled for direct analysis, even the uppermost mantle is inaccessible for direct analysis. According to Figure 10.7, the upper mantle should consist of rocks with a density of ~3,400 kg m^{-3} and a p-wave velocity of ~8 km s^{-1}. Small rock fragments with these characteristics are occasionally brought to the surface from the mantle as nodules or xenoliths by ascending magmas, particularly in ocean islands, and by diamond-bearing kimberlite extrusions in ancient continental regions, as in South Africa. As diamond is a high pressure form of carbon, kimberlites must originate from at least 150 kilometre deep in the mantle. Much larger samples of such rocks are also exposed in ophiolites, which comprises both ocean crust and upper mantle rocks that have been upthrust in zones of former plate convergence.

The direct samples from the upper mantle (both nodules and ophiolites) are invariably Mg- and Fe-rich silicates, collectively called peridotites (Brown and Mussett, 1993). Peridotites represent an extensive category of ultramafic rocks consisting of olivine (Mg, Fe)$_2$Si$_2$O$_6$, commonly > 50 per cent, clinopyroxne (Ca, Mg, Fe)$_2$Si$_2$O$_6$, and an aluminous phase, either plagioclase (CaAl$_2$Si$_2$O$_8$), spinel (MgAlSiO$_4$), or garnet (Mg, Fe)Ca$_3$Al$_2$Si$_2$O$_{12}$ depending on the confining pressure. These are the same minerals as found in meteorites. The seven major elements in meteorites and the Earth dictate the structure of the above minerals in the mantle. The less abundant elements have to fit to the extent possible, either in minor (accessory) minerals, or in atomic substitution in the major mineral phases.

For peridotites to be the principal rocks in the upper mantle, they must be capable of producing a large quantity of MORB on partial melting. Many experimental studies have shown that basalt is indeed the partial melt of mantle peridotites. In fact, the compositions of nodules in basalts indicate two types of peridotites; one with all the main and essential minerals and another with only olivine and orthopyroxene. The former is called lherzolite or fertile peridotite (before any basalt extraction), and the other, harzburgite, to represent peridotites (depleted peridotite) after basalt extraction.

Eclogite is another rock found in the upper mantle. It is a dense, high pressure metamorphic rock with a bulk composition resembling basalt. Mineralogically, eclogites contain roughly equal parts of an aluminous pyroxene (Omphacite, which is a clinopyroxene with some CaMg replaced by NaAl) and garnet.

10.9   Earth as a Large Geochemical System

A geochemical system, in general, is defined as that portion of the universe selected for focused study for either scientific understanding, or practical applications (Richardson and McSween, 1989). The scope of the study determines the dimensions of the system. A system can be isolated, closed, or open. An isolated system exchanges neither energy nor matter with its surroundings; a closed system exchanges energy, but not matter; whereas, an open system exchanges both. Most natural systems, especially large ones, are either closed or open. In the specific example of the Earth, it has remained closed for all practical purposes since its formation, but not so are its stratified layers. As we saw earlier, the latter have been continually exchanging matter and also energy in some cases, as a result of solid state convection and associated plate tectonics, which are believed to have been operational for at least the last half of the earth's history, if not earlier in some other form. If the rate of exchange of matter between subsystems is very low, they may be considered closed on smaller time scales. Thus, time is an important consideration in the choice of the subsystems.

Box models are commonly used to describe a system made up of many different, but interlinked subsystems or reservoirs. Figure 10.9 shows such a model for the Earth, in which the boxes refer to the chemical systems in the Earth, and the arrows between boxes the pathways connecting them. A single arrow indicates one directional transfer, a double-headed arrow bidirectional transfers. This diagram is a simplified version of the most comprehensive geochemical models for the Earth, called the Geochemical Earth Reference Model (GERM) (Staudigel et al., 1998) and is similar to the more familiar geophysical model, known as the Preliminary Reference Earth Model (PREM) (Dziewonski and Anderson, 1981). The GERM divides the Earth into a comprehensive set of geochemical reservoirs, and provides an internally consistent set of data describing these reservoirs and isotopic composition and quantitative estimates for the chemical fluxes between these reservoirs over relevant time scales. The figure includes three reservoirs in addition to those mentioned already. These are the ocean, atmosphere, and biospshere. The uppermost box with dashed outlines represents external inputs (both continuous and sporadic) to the Earth. Comet and asteroid impacts do not constitute significant addition of matter, but their kinetic energy can seriously disturb the Earth system, especially the biosphere, at least temporarily. We recall that the continental crust and the underlying lithosphere are long-lived features, whereas, the oceanic crust and the underlying lithosphere are transient with an average lifetime of ~100 million years.

Many of the linkages shown in this diagram may be easily recognized. A few additional explanations are in order. The core, presumably formed even during the accretion of the Earth, consisted entirely of molten iron with minor elements dissolved in it. With the passage of time a solid inner core began to crystallize out (CR). Many siderophile elements would have preferentially partitioned into the melt phase (EX). The primary processes are the generation of voluminous oceanic crust at ocean ridges (RD), its alteration by hydrothermal circulation of sea water (HY), and finally its elimination at subduction zones (SD). Gases and fluids generated by this volcanism

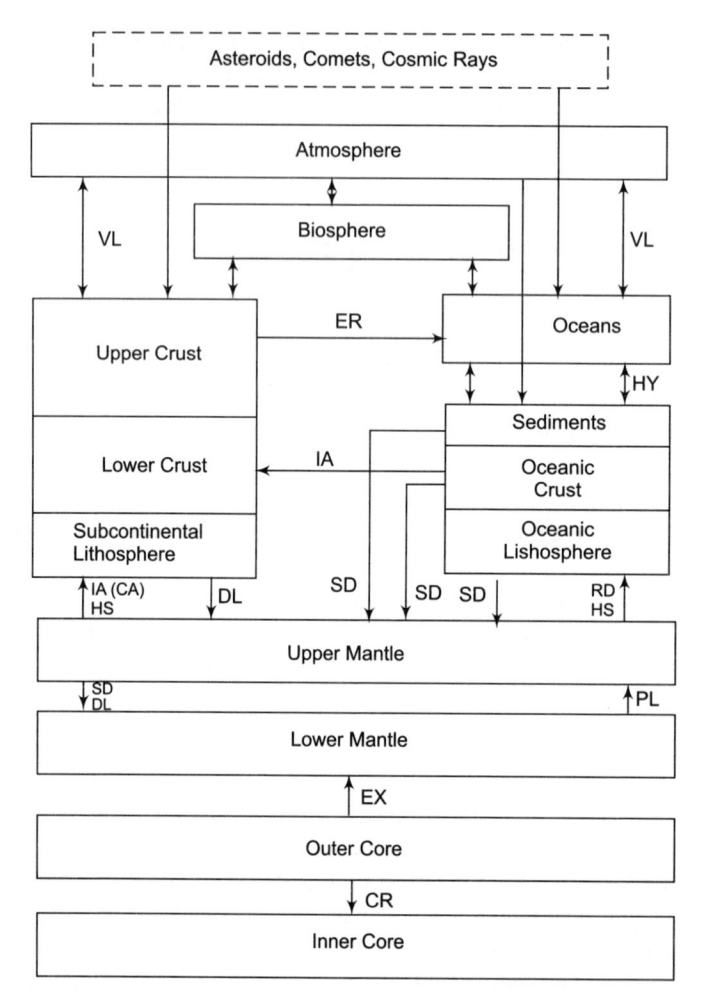

Figure 10.9 Schematic diagram of the geochemical reservoirs and interactions between them. CR, crystallization; EX, exsolution; SD, subduction; RD, ocean ridges; HS, hot spots; Pl, plumes; DL, delamination; IA, island arcs; CA, continental arcs; VL, volcanism; HY, hydrothermal circulation; ER, erosion.

transfer incompatible and volatile elements, including radiogenic helium (He), and argon (Ar), to the oceans and atmosphere (VL). A relatively small quantity of oceanic crust is generated by hotspots (HS). Continental crust is created by complex processes in subduction zones as island arcs (IA) and continental arcs (CA). Continental crust exposed over ocean is eroded (ER) and deposited as a thin layer of sediments over the spreading oceanic crust. Some of these sediments are carried by the subducting plate for incorporation into the subduction zone volcanic rocks. However, all the known contributions to the continental crust are considerably more mafic (basaltic) than its present silicic composition. One explanation for the silicic composition is that the magmas from the mantle associated with subduction zone volcanics, flood basalts, and hot spots intrude the continental crust, and in the presence of water produce silicic (granitic) magmas. These magmas rise into the upper crust, making the upper crust more silicic and the lower crust more mafic. Subsequently, the dense mafic dense rocks of the lower crust are returned to the mantle by delamination (DL). The net result is that continental crust becomes more silicic.

The coupling and feedbacks between the biosphere and the ocean, atmosphere and upper crust is the basis of Lovelock's insight (Lovelock, 1979; 1988) that life affects the global environment and the surficial processes on the Earth. He did not mean that the Earth is a living organism, but behaves like one in some respects. Lovelock laid the foundation for the development of 'systems thinking' in Earth science. Simply put, systems thinking means that the whole could be greater than the sum of its individual parts by way of displaying 'emergent properties' that can never be understood from the traditional reductionist approach (Langmuir and Broecker, 2012).

10.10 ELEMENTAL CHEMISTRY OF MID-OCEAN-RIDGE BASALTS, OCEAN-ISLAND BASALTS AND CONTINENTAL CRUST

It is far beyond the scope of this book to consider the complex processes leading to the eruption of magma to the surface at divergent plate boundaries, convergent plate boundaries, and in mid-plate locations. We restrict ourselves to broad inferences from the distribution of incompatible elements in rocks derived directly or indirectly by partial melting in the mantle. We recall that incompatible elements are those with a partition or distribution coefficient, $D < 1$. In the context of mantle melting, this means that if a region of the mantle partially melts, the melt will be preferentially enriched in these incompatible elements, leaving the parental or source region correspondingly depleted in them. The equation of partial melting shows for highly incompatible elements with $D \ll 1$, the enrichment for all of them in the melt will be by the same factor, $(1/F)$, where F is the degree of partial melting (amount of melt/amount of the initial solid). Thus, the concentrations of more incompatible elements in the source region prior to melting will all be lower than their concentrations in the melt by a factor of $(1/F)$.

MORB and OIB are direct mantle melts, whereas, continental crust is believed to be a complex body made up of many melt extracts. Figure 10.10 presents a plot of the concentrations of many major and trace elements relative to the primitive mantle (BSE) and, in increasing order, of compatibility in average continental crust, average oceanic crust (MORB), and three selected OIBs: OIB 1 from Hawaii, OIB 2 from Tribuai, Austral Islands, and OIB 3 from Tristan (Hofmann, 1997). Hofmann focused on these ocean basalts rather than mantle xenoliths or island arc volcanics, because oceanic basalts represent relatively large volumes of mantle and are least likely to be contaminated during melt transport from the source region to the surface through the intervening crust. This plot illustrates a number of important points, as brought out by Davies (1999). The plot shows that incompatible elements are enriched to various degrees in all the five rocks relative to the primitive mantle (line corresponding to ratio = 1). The melt fraction, F, for MORB is estimated to be 0.1 on the average. This means that the concentrations of incompatible elements in the MORB source region is ten times lower than in MORB. It can be seen from the figure that the MORB source now has lower than primitive mantle concentrations of the more incompatible elements. In other words, the MORB source is depleted of incompatible elements. This implies that this batch of MORB has been formed by melting of a mantle source region that has been melt-depleted at least once previously. This can also be inferred from the relatively larger depletion of the most highly incompatible elements, like barium Rb, Ba and Th, compared to the moderately incompatible elements, like Sm and Hf, in the MORB.

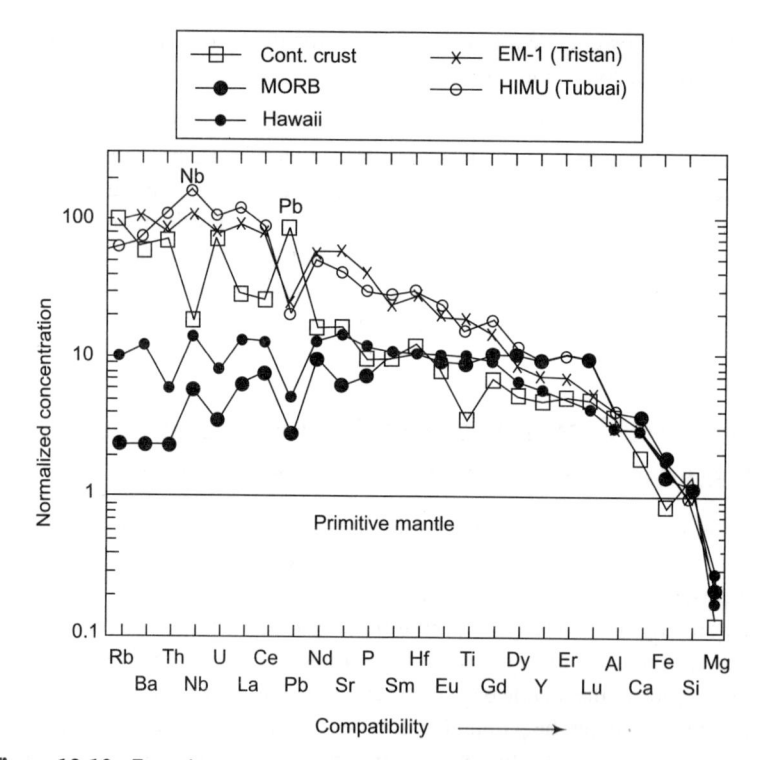

Figure 10.10 Trace element concentrations in mantle-derived rocks, normalized to estimated concentrations in the PM or BSE (Hofmann, 1997).

The three OIBs are believed to be derived by the melting of mantle plumes from the deeper mantle to a degree less than under mid-ocean ridges (MORB). Thus, they must be more enriched in incompatible elements than MORB, which they indeed are as seen from the figure. But, whereas, the enrichment factor in OIB 1 is barely ~10, it is much higher, ~100 for OIB 2 and OIB 3 relative to primitive levels. However, the similarities in their Nb and Pb anomalies show that despite the widely different levels of enrichment, the three OIBs have fundamentally similar trace element patterns. If the difference in enrichment levels is only due to difference in the melting process, the melt fraction should be as high as 10 percent for OIB 1 and as small as ~1 per cent for OIB 2 and 3. The possibility that the sources of OIB 2 and 3 are enriched relative to primitive levels cannot be ruled out. It seems reasonable to infer that plume sources are less depleted on average than the MORB source.

Continental crust differs from oceanic crust in important respects, as we noted earlier. Oceanic crust is relatively homogeneous with a basaltic composition and young age (~100 million years). But continental crust is very complex, heterogeneous on all scales with a bulk andesitic composition, and populated by a wide variety of rocks of different ages. So it is, therefore, surprising that the incompatible element abundance patterns of the continental crust and the two enriched OIBs would have been indistinguishable in the figure, but for their Nb and Pb anomalies being opposite to each other. Had it not been for the difference in the sense of the Nb and Pb anomalies, the continental crust could be approximated quite well as a result of a single stage partial melting of the primitive mantle to about ˋ1 per cent ($F = 0.01$). Although such a simple way to produce continental crust is not plausible, it does suggest that the mechanism for the enrichment of incompatible elements in the continental crust is still partial melting, but under conditions different from those in the generation of MORB and OIB. As we noted earlier, subduction zone volcanism differs from ridge and hot spot volcanism in that the former depends crucially on the release of water from hydrated minerals in the subducting oceanic crust and its rise into the overlying mantle wedge to cause its melting at a lower temperature. The resulting incompatible-element-loaded melt rises to form island arcs. Nb is more compatible and Pb is less compatible in such wet silicate melts than in dry silicate magmas. Ti is believed to be more compatible because of the occurrence of ilmenite ($FeTiO_3$) in this environment. It is inferred from this and other evidence, that island arc volcanism contributes significantly to the growth of the continental crust at least after the initiation of plate tectonics.

With a mass of only 0.6 per cent of the Earth's mass, the continental crust has extracted over the Earth's history up to 70 per cent of the Earth's budget of highly incompatible elements. As only the mantle could have supplied so much material, it should have been depleted in the elements that are enriched in the continental crust.

This raises the question whether the entire mantle or only a part of it contributed to the incompatible element enrichment in the continental crust. In the first case, the entire mantle will be depleted to some degree. In the latter case, only a fractional part of the mantle will be depleted, but to a larger extent, in incompatible elements relative to primitive levels. Depletion of the entire mantle will be facilitated by stirring the whole mantle pot so that all material in the mantle is brought up to the shallow melting region. This is termed whole mantle convective circulation. Depletion of only a fraction of the mantle will mean that convection in that part of the mantle only transports material to the shallow melting region. Available geochemical evidence shows that the upper part of the mantle is substantially and uniformly depleted, whereas, the lower mantle is much less depleted, if not even enriched in incompatible elements. This geochemical evidence favours convection of the mantle in two chemically isolated layers or tiers. On the basis of seismic tomographic evidence for a significant exchange of mass between the upper and lower mantle, geophysicists advocate whole mantle convection. Reconciliation of these physical and chemical observations is a major issue in modern research.

Bearing also on the style of mantle convection is the existence and origin of mantle plumes or hot spots believed to be the source of ocean island volcanism. Again, there are two contending models (Figure 10.11): one favouring origin of mantle plumes in a source at the core-mantle boundary and the other favouring shallower plume sources within the upper mantle. A deep origin for mantle plumes would support whole-mantle convection and also recycling of ocean crust into the deeper mantle to explain the enrichment of any incompatible elements and the presence of isotopically distinct components within the lower mantle (Hofmann, 1997).

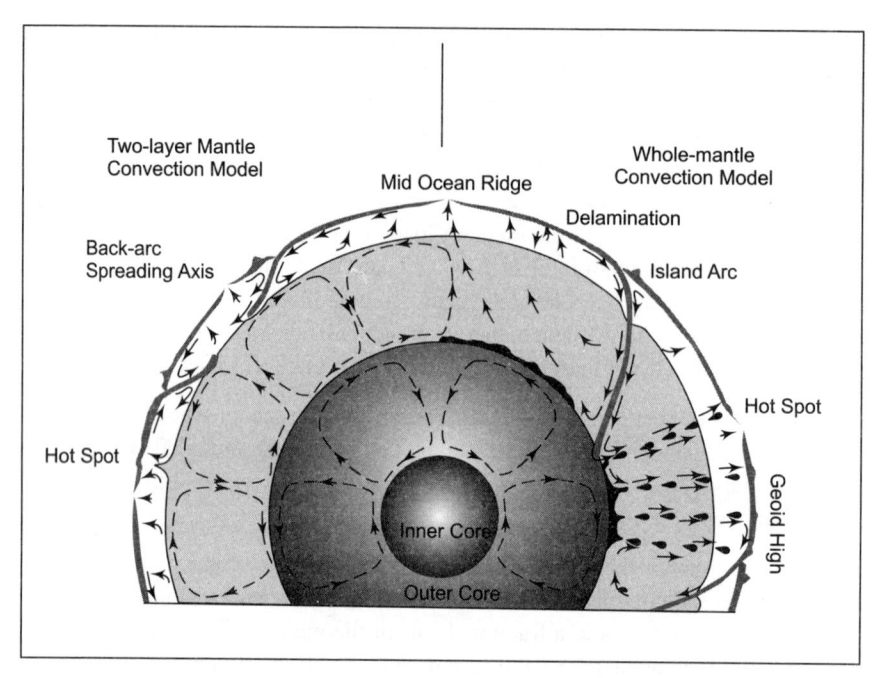

Figure 10.11 Two models of the mantle convection; one involving separate convection of the upper and lower mantles (left), and the other involving whole mantle convection (right) (Philpotts, 1990).

11 Chronology of Earth History

> Geology is the study of complex natural experiments conducted on a large scale in both time and space. The experiments are neither reversible nor repeatable. They cannot be directly observed; but they must be reconstructed historically.
>
> **S A Schumm**
>
> We view samples of the accessible universe as a library of experiments that have already been done (at different times and sites). Our art must be to select those materials which have a persistent memory and which together can bring testimony to the natural experiment of interest.
>
> **G J Wasserburg**

11.1 Introduction

Simply stated, the present Earth is very different from the early Earth. The main objective of geology is to reconstruct the time sequence and nature of the key or major processes that shaped the Earth over 4.5 Gy. Schumm (1991) cautions that this task will be quite formidable by quoting a metaphorical statement of Pretorius in a different context, as follows:

> It is the nature of the history of the earth that a geologist has available to him only partial information. Occasional lines from disconnected paragraphs in obscuranist chapters are what can be read. Violence in the handling of the book through time has caused many of these chapters to be ripped and reassembled out of context. That the gist of early chapters can be deciphered at all is a credit to the perseverance and imagination not always associated with other sciences. The geologist operates at all times in an environment characterized by a high degree of uncertainty and ornamented with end products which are the outcomes of the interactions of many complex variables. He sees only the end, and has to induce the processes and responses that filled the time since the beginning.

To extend this metaphor further, some of the missing pages of the book of genesis of the Earth may have to be found in the libraries of experiments carried out on other planets.

In this chapter we will consider how radiogenic isotope geochemistry has provided firm constraints on the timing of some major or large scale events in Earth's history. Radiogenic isotope geochemistry, as we have noted many times, rests on two powerful tenets, namely:
1. The dependence on radioactive decay introduces the possibility of referring the history to an absolute time scale; and
2. The isotopic composition of an element is insensitive to chemical fractionation caused by equilibrium partial melting or fractional crystallization.

Thus, when a partial melt from the mantle rises to the surface, it extrudes as basaltic lava and delivers an implicit chemical (and an explicit isotopic) message from the mantle to the geochemist (Hofmann, 1997), thereby enabling him or her to sample the deep interior of the Earth much as geophysicists do. The following equations based on the two principles specify the isotopic evolution, in general, in a parent system A, formed at time t in the past, and its chemical derivatives, B and C at a later time, t_1. The parent system is described by the fundamental age equation as:

$$^A\left(\frac{d}{d'}\right) = {}^A\left(\frac{d}{d'}\right)_t + {}^A\left(\frac{p}{d'}\right)[\exp(\lambda t) - 1] \qquad (11.1)$$

where, $^A(d/d')$ and $^A(p/d')$ are the daughter isotope and parent-daughter composition ratios, respectively, at present. The decay constant is known independently, leaving t and $^A(d/d')_t$ as two unknowns. The isotopic evolution of B and C since t_1 will be given by:

$$^B\left(\frac{d}{d'}\right) = {}^B\left(\frac{d}{d'}\right)_{t1} + {}^B\left(\frac{p}{d'}\right)[\exp(\lambda t_1) - 1] \qquad (11.2)$$

$$^C\left(\frac{d}{d'}\right) = {}^C\left(\frac{d}{d'}\right)_{t1} + {}^C\left(\frac{p}{d'}\right)[\exp(\lambda t_1) - 1] \qquad (11.3)$$

Because chemical fractionation discriminates only elements, but not their isotopic ratios,

$$^A\left(\frac{p}{d'}\right)_{t1} \neq {}^B\left(\frac{p}{d'}\right)_{t1} \neq {}^C\left(\frac{p}{d'}\right)_{t1}$$

$$^A\left(\frac{d}{d'}\right)_{t1} = {}^B\left(\frac{d}{d'}\right)_{t1} = {}^C\left(\frac{d}{d'}\right)_{t1} \qquad (11.4)$$

Strictly speaking, isotopes of an element will undergo a small mass-dependent fractionation during chemical and physical processes. But measurement protocols for radiogenic isotopes correct for such mass discrimination. Stated in words, two or more subsystems derived by chemical differentiation of a parent source will have different parent-daughter ratios, but the same daughter isotope composition as the parent source immediately after their formation. They will become isotopically distinct only with the passage of time. The time evolution of radiogenic isotope ratios in the three systems is shown in Figure 11.1 for radioactive nuclides that do not survive to the present and those that do. These are called short-lived and long-lived isotopes, respectively. As short-lived isotopes do not survive to the present, the variation in their daughter isotope abundance must be tracked relative to time measured forward.

In all radiogenic isotope-based studies on the evolution of the Earth, it is assumed that Earth inherited at its birth the isotopic composition of radiogenic elements exactly as in meteorites, which we called primordial in Chapter 9. Although volatile elements may be depleted in the Earth relative to meteorites, refractory elements are believed to occur in the same proportions in both.

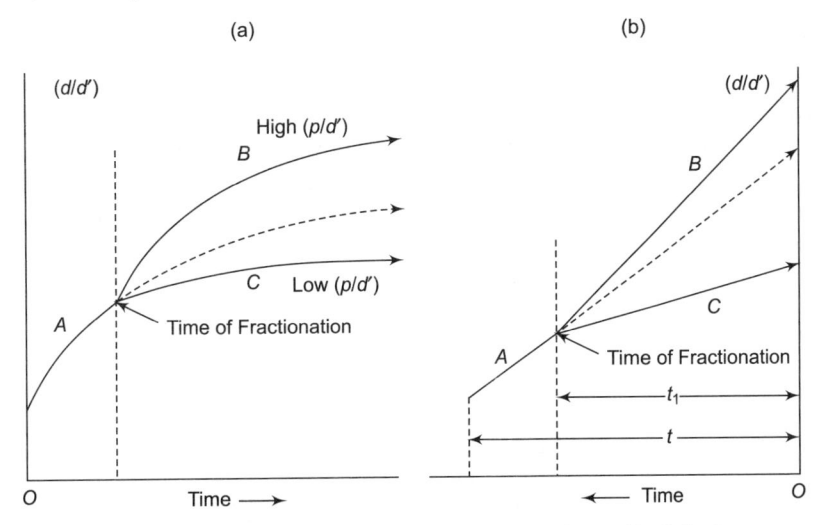

Figure 11.1 Daughter isotope evolution in System A and two of its derivatives, B and C for short-lived parents (a), and long-lived parents (b).

11.2 EARLY SIDEROPHILE-LITHOPHILE SEGREGATION AND TIMING OF CORE FORMATION

The parent-daughter pair best suited to date core separation is one in which the parent and daughter elements are strongly fractioned by metal-silicate segregation. As the core is believed to have formed shortly after earth accretion, it is best tested if the parent radioactive isotope is, in addition, short-lived relative to the age of the Earth. In that case, even the mere detection of the former presence of such an extinct isotope in any of the Earth layers would be clinching.

The decay of one of the isotopes of the lithophile element hafnium, ^{182}Hf to one of the isotopes of the siderophile element tungsten, ^{182}W with a half-life of 9 million years meets these criteria. As both Hf and W are refractory elements, their relative abundances in the unfractionated Earth would have been the same as in primitive meteorites. As and when the core separated, W will be strongly fractionated from Hf, with siderophile W mostly entering the core and the lithophile Hf remaining entirely in the rocky silicate layer. However, if the separation took place before all the ^{182}Hf decayed away, that is, within 50 million years, the mantle with the highest Hf/W ratio will produce a much higher proportion of ^{182}W relative to a non-radiogenic isotope of tungsten, like ^{184}W, than the bulk Earth (or meteorites) and even higher than the core as shown by Curve B in Figure 11.1. After about five half-lives, ^{182}Hf will be essentially extinct in the core and mantle, and their ^{182}W/^{184}W ratios will level off at different values, as shown in Figure 11.1a. The earlier the core separated after earth accretion, the more extreme would be the difference in W isotope ratios. The steady state ratio of the unstable ^{182}Hf isotope to a stable isotope of Hf even in the freshly isolated solar nebular cloud will be extremely small, ~ 10^{-4}, unless boosted by, so to speak, a 'last minute' spike, as discussed in Chapter 8. Differences in W isotope ratios will also be of this order. Measurement of such small differences in the isotope ratios of Hf and W were possible only by 2002 (Yin et al., 2002; Schoenberg et al., 2002). A measurable difference in the right direction between a sample of the core (Curve C) and that of the mantle (Curve B) in Figure 11.1a would be unambiguous evidence for core separation in the very early Earth. As no sample of the core ever reaches the surface, enrichment of the W isotopic ratio in the mantle, relative to carbonaceous chondrites, is the best evidence we have for early core separation (Figure 11.2). The magnitude of this enrichment (about 2 parts in 10,000) has been interpreted as formation of the core in the first 30 million years of the Solar System history. This assumes that core segregation was a single step-event. It is not necessary that a piece of the mantle be analysed for its W isotopes. After the total decay of ^{182}Hf, any tungsten-containing derivative from the mantle would do; for example, even the fine tungsten coil in an old-fashioned incandescent electric bulb.

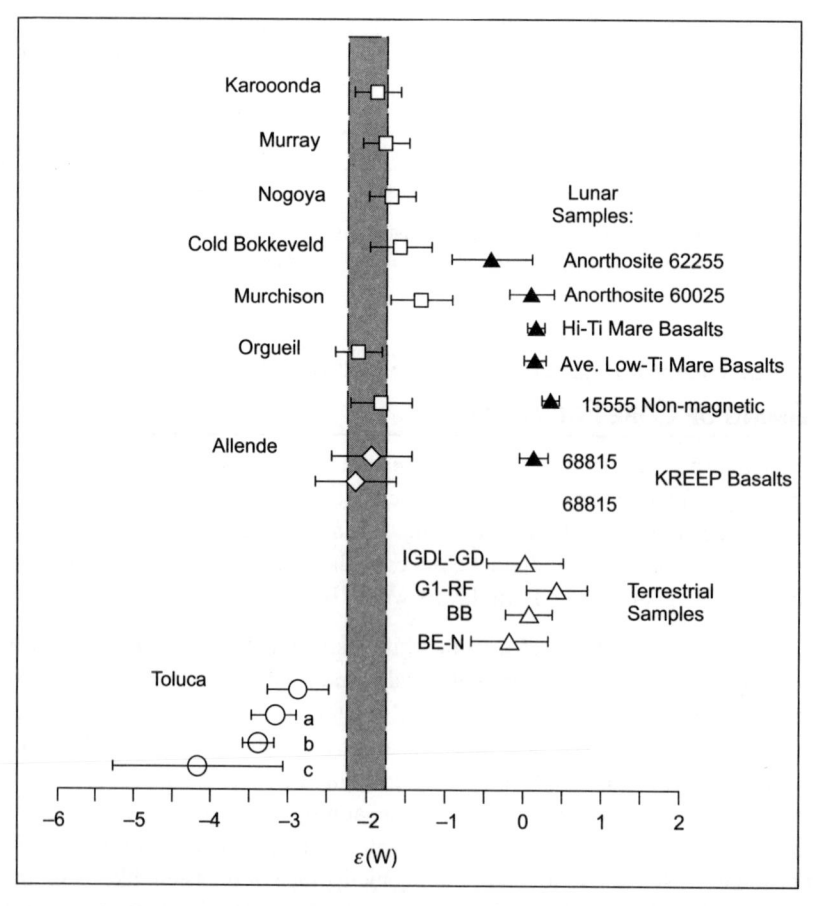

Figure 11.2 Tungsten isotope ratios measured in meteorites and terrestrial and lunar materials relative to a terrestrial standard (from White, 2013, based on data from Kleine et al., 2002; Yin et al., 2002; Touboul et al., 2007, 2009).

Because the lithophile uranium (U) contrasts with chalcophile lead (Pb), the U-Pb system can, in principle, be used to time core generation. But the fractionations of U and Pb during the long history of the Earth preclude the precise timing of core segregation. However, studies of U-Pb fractionation are consistent with core separation within 70–150 Ma after the formation of the Solar System.

11.3 Early Lithophile-Atmophile Separation and Timing of the Primitive Atmosphere

The elements and compounds that remained uncondensed in the inner solar nebula are presumed to have been blown off by the solar wind from the inner solar system during the T-Tauri phase of the infant Sun, leaving the Earth with only traces of uncondensed solar gases that were trapped in the condensed matter. Volatile elements and compounds trapped in the interior of the Earth must have been expelled or outgassed when the primitive Earth melted extensively for metal to separate from silicates. The gravitational retention of the outgassed components as a gaseous envelope around the Earth must represent its first 'indigenous' atmosphere. Water vapour from this envelope must have then condensed to form the hydrosphere or oceans. Other reactive components may have been subsequently removed from the primitive atmosphere, but the unreactive, or inert components, like the rare gases, will accumulate with time in the atmosphere. Rare gases found in meteorites and the Earth are a variable and complex mixture of many isotopically distinct components, details of which may be found in Ozima and Podosek (1983), Ozima (1994), and McDougall and Honda (1998). We will restrict ourselves only to radiogenic components in rare gases to infer time scales of causative processes.

Table 11.1 Volume Fraction of Rare Gases in the Present Atmosphere

Element	Isotopic Masses	Volume Fraction
Helium	3, 4	5×10^{-6}
Neon	20, 21, 22	2×10^{-5}
Argon	36, 38, 40	1×10^{-2}
Krypton	78, 80, 82, 83, 84, 86	1×10^{-6}
Xenon	124, 126, 128, 129, 130, 131, 132, 134, 136	1×10^{-7}
Nitrogen	14, 15	7.8×10^{-1}
Oxygen	16, 17, 18	2.1×10^{-1}

The volume fractions of the five rare gases in the present atmosphere are given in Table 11.1, together with their isotopic makeup. Whereas, helium (He), neon(Ne), krypton (Kr), and xenon (Xe) are indeed very rare in the atmosphere, argon (Ar) is far more abundant. It is the third largest [after nitrogen and oxygen] constituent of the atmosphere with a volume fraction of ~1 per cent, the reason for which is given later. Of these five gases, only He, Ar, and Xe contain radiogenic components. ^{3}He is a primordial nuclide, and inherited by the Eearth from the solar nebula, ^{4}He is also produced by the radioactive decay of ^{235}U, ^{238}U, and ^{232}Th, and, hence, will increase with time in objects containing U and Th. ^{36}Ar and ^{38}Ar are both primordial and non-radiogenic, whereas, ^{40}Ar is the decay product of ^{40}K. Of the nine isotopes of xenon, ^{129}Xe is radiogenic from the beta decay of I^{129} with a very short half-life of 16 Ma. Spontaneous fission of ^{238}U and the transuranic isotope ^{244}Pu produces heavier isotopes of Xe, which we will not consider further. Table 11.2 gives the radiogenic isotope composition of helium, argon and xenon in different Earth reservoirs, together with their primordial values as inferred from meteorite studies. These ratios follow the regular convention of placing the radiogenic isotope in the numerator and the non-radiogenic isotope in the denominator. The reciprocal of this ratio, ^{3}He/^{4}He, was introduced by Harmon Craig and his coworkers in the University of California, San Diego to focus on primordial helium in earth reservoirs (Lupton and Craig, 1975), and is now commonly used in literature with its atmospheric value of 1.4×10^{-6} taken as a reference composition.

Table 11.2 Radiogenic Isotope Composition of Helium, Argon, and Xenon in Various Earth Reservoirs

Isotope Ratio	Primordial	Atmosphere	Mid-Ocean-Ridge Basalts	Ocean-Island Basalts
$\dfrac{^{4}He}{^{3}He}$	10^{4}	714×10^{3}	$\sim 90 \times 10^{3}$	$13–130 \times 10^{3}$
$\dfrac{^{40}Ar}{^{36}Ar}$	10^{-4}	296	up to 30×10^{3}	up to 13×10^{3}
$\dfrac{^{129}Xe}{^{130}Xe}$	6.3	6.5	7.65	~7.0

Any outgassing event will separate virtually all the atmophile daughter elements from their lithopile parents in the mantle. Because He, Ar, and Xe will eventually end up in the atmosphere, the radiogenic isotope composition of these elements in the atmosphere will reflect the sum total of all outgassing events throughout Earth history. The extremely short half-life of ^{129}I is ideally suited to distinguish between an early major outgassing event, as opposed to the gradual or intermittent outgassing over the whole of Earth history. Figure 11.1a depicts the scenario of the atmosphere forming as the result of a major outgassing event at time t_1 early in the history of the earth while ^{129}I was still extant, but greatly diminished. The fractionation of Xe into the atmosphere and the preferential retention of iodine in the mantle would result in a greater proportion of ^{129}Xe in the degassed mantle than in the atmosphere. Thus, an important clue for early mantle outgassing is a measurable excess of ^{129}Xe in the mantle relative to the atmosphere. Claude Allegre and his coworkes in Paris showed in 1983 (Allegre et al., 1983) that the Xe ratio in Mid-Ocean-Ridge Basalts (MORB) is slightly higher than in the atmosphere which, in turn, is slightly higher than in primordial Xe, as given in the Table 11.2. If the outgassing occurred after the exhaustion of ^{129}I (~100 My), all Earth reservoirs will have identical xenon ratios. The small Xe isotopic anomaly has been modelled to constrain the time of the earliest outgassing to within ~30 My of Solar System formation, surprisingly similar to the time of core formation based on Hf-W systematics. This finding exemplifies what Wetherill told the author in 1976 in a different context that 'an effect need not be spectacular to be significant'.

Whereas, ^{129}I with its very short half-life of 16 Ma became extinct within 100 Ma of earth history, ^{235}U, ^{238}U, ^{232}Th and ^{40}K, with much longer half-lives, are still extant and generate ^{4}He and ^{40}Ar, respectively. ^{4}He and ^{40}Ar abundances can track mantle outgassing due to volcanism at mid-ocean ridges and midplate ocean islands subsequent to the early degassing event indicated by xenon isotopes. Helium is not useful in this respect as helium outgassed from the mantle will not be gravitationally bound to the atmosphere, but will partially escape from it into space: lighter ^{3}He escaping at a slightly faster rate than the heavier ^{4}He. This explains the very low abundance of helium in the atmosphere despite the copious production of ^{4}He from the radioactive decay of uranium and thorium.

The equilibrium ^{4}He/^{3}He ratio in the present atmosphere is 7,14,000. In MORBs this ratio is reduced to 90,000. In most Ocean-Island Basalts (OIBs) this ratio is varies from 1,30,000 to 13,000. The lower ratios in ocean volcanics are taken to imply that the mantle has a higher proportion of primordial helium, which leaks through volcanism in the mid-ocean ridges and ocean islands.

As mentioned before, helium, and to some extent neon, are not gravitationally bound to the atmosphere, but the heavier rare gases argon, krypton and xenon are quantitatively retained, and accumulate there if outgassed from the interior of the Earth. The accumulation of radiogenic ^{40}Ar from the decay of ^{40}K outgassed by the Earth explains the exceptionally high concentration of argon in the atmosphere and its almost completely radiogenic (99.56 percent) composition. ^{36}Ar is primordial and non-radiogenic, but constitutes only 0.4 per cent of the total, as given by the ^{40}Ar/^{36}Ar ratio of 296 in the atmosphere. This ratio is much higher, up to 30,000 in MORB and at least 13,000 in OIBs. These high ratios are taken to imply that atmosphere represents the bulk of the primordial argon inherited by the Earth.

Table 11.2 reveals a glaring contrast between radiogenic isotope ratios of helium (e.g. ^{4}He/^{3}He) and helium (e.g. ^{40}Ar/^{36}Ar) in the mantle and atmosphere. The Ar ratio in the atmosphere (296) is very much smaller than MORB and OIB values. It is just the opposite in helium, where the atmospheric ratio far exceeds MORB and OIB data. This is easily explained as due to the quantitative retention of argon in the atmosphere and continuous leakage of helium from it throughout Earth history. The very high helium ratios arise from the continuous depletion of ^{3}He, but replenishment of ^{4}He from radioactive decay of uranium and thorium in the continental crust and mantle.

Earth inherited only traces of primordial argon. As ^{40}Ar is very low in primordial argon (^{40}Ar/^{36}Ar = 10^{-4}), it was realized quite early (Allegre et al., 1996) that the ^{40}Ar inventory in the Earth at present is entirely radiogenic, generated over 4.5 Gy from an estimated potassium (K) content of ~240 ppm in the silicate earth. This total ^{40}Ar must now be distributed between the atmosphere, continental crust, and mantle, assuming that the metallic core does not contain any Ar. The quantitative ^{40}Ar balance between the three reservoirs is given in Table 11.3 (Davies, 1999).

Table 11.3 Argon Abundances in the Major Earth Reservoirs

	Quantity	Per cent
Total generated in 4.5 Ga (calculated)	375×10^{16} mol	100
In the atmosphere (measured)	165×10^{16} mol	44
In continental crust (estimated)	13×10^{16} mol	4
In the entire mantle (balance)	194×10^{16} mol	52

Taken at its face value, this result indicates that about 50 percent of all radiogenic argon generated during the whole history of the Earth resides in the mantle and the Earth is, therefore, only about 50 per cent degassed. But, when it is estimated that the upper mantle at present cannot account for more than 1 per cent of this amount, it leads to the startling conclusion that the lower mantle is the repository of so much radiogenic argon. This large apparent disparity between the upper and lower mantle is usually explained by postulating that the upper mantle is extensively degassed, and that its original Ar content is now in the atmosphere, whereas, the lower mantle is not outgassed and contains the original primordial and radiogenic argon (and potassium content). This requires, in turn, that the lower and upper mantles have remained chemically isolated for the major part of Earth history, lending strong support to the two-layer convection model of the mantle. This geochemical evidence cannot be reconciled with geophysical evidence, which favours whole mantle convection, unless the assumed K content of the silicate Earth is a gross overestimate by a factor of almost two (Albarede,1998; Trieloff et al., 2003).

In summary, the following broad aspects of the atmosphere and mantle are suggested by radiogenic isotopes of helium, argon and xenon: (1) The present atmosphere is the product of degassing of the Earth's interior; (2) A substantial portion of the mantle has been degassed; and (3) Most of the degassing occurred very early in Earth's history. Other rare gas components in the atmosphere are not easily explained, rendering the atmosphere the least understood of Earth's layers (Langmuir and Broecker, 2012).

11.4 Lithophile-Lithophile Separation and Timing of the Early Crust

The separation of the core and the outgassing from the Earth in the first few tens of millions of years of Earth formation must be the result of planet-wide melting to form a magma ocean at that time. This magma ocean must have subsequently solidified, at least in its upper parts, to form a sufficiently cool and solid crust for water vapour to condense into a primitive ocean. The obvious question then is when such a crust formed, if at all it did, and if any of it survives to this day for sampling and analysis. This has inspired a spirited quest for the oldest rocks on the Earth and the type of information that can be extracted from them. While rubidium-strontium (Rb-Sr), samarium-neodymium (Sm-Nd) and U-Pb dating of whole rocks and their constituent minerals have been used to recognize old rocks, the U-Pb dating of the small accessory mineral in felsic rocks, zircon is the preferred method for the following reasons. First, the zircon ($ZrSiO_4$) structure readily accepts U because it has the same charge (+4) and similar size (89 picometre, pm) as Zr (+4 and 92 pm, respectively), but strongly rejects Pb with its different charge and radius (+2 and 119 pm, respectively). In fact, even micron size regions of a single zircon grain have sufficient U to be individually dated *in situ* by specially designed ion probes, such as the Sensitive High-Resolution Ion Microprobe (SHRIMP) developed by W. Compston and his coworkers at the Australian National University (Compston et al., 1984); Second, the relatively short half-lives of both ^{235}U and ^{238}U facilitate rapid accumulation of radiogenic ^{207}Pb and ^{206}Pb for precise age determination; Third, zircon is a chemically and mechanically robust mineral, which resists weathering, alteration, chemical breakdown, and abrasion during stream transport from their source rocks to their final incorporation in sedimentary rocks; Fourth, discordance between the two U-Pb ages resulting from open system effects (Chapter 5) can be corrected using the concordia-discordia method. Thus, U-Pb zircon dating has become the 'gold standard' for determining the ages of the oldest rocks and, hence, continental growth (White, 2013).

Geological time is classified into four major subdivisions, namely: (1) the Hadean, spanning the time interval from the formation of the Earth about 4.6 Ga to 3.8 Ga; (2) the Archean from 3.8 Ga to 2.5 Ga; (3)The Proterozoic from 2.5 Ga to 600 Ma; and (4) the Phanerozoic from 600 Ma to the present. Rocks older than 3.8 Ga are extremely rare, but Archean rocks in the age interval 3.8 to 2.5 Ga are relatively common, and are preserved in what are known as cratons or shields in many major continents, as shown in Figure 11.3 (Rollinson, 2007). Archean cratons contain three different types of lithological association, namely, Archean greenstone belts, late Archean sedimentary basins, and Archean granite-gneiss terrains. The first two are believed to represent the upper Archean continental crust and the last the middle and lower Archean continental crust. The relationships between these three associations are decipherable in only few locations, one of which is the Kapuskasing structural zone in the southern part of the Superior Province in Canada (Percival and West, 1994). There are rock types that are unique to the Archean, implying that the Archean Earth must have somehow been different. Examples of these rock types include the ultramafic lavas, komatiites, sedimentary Banded Iron Formations (BIF) found in green stone belts, and calcic anorthosites (De Wit and Ashwal, 1997).

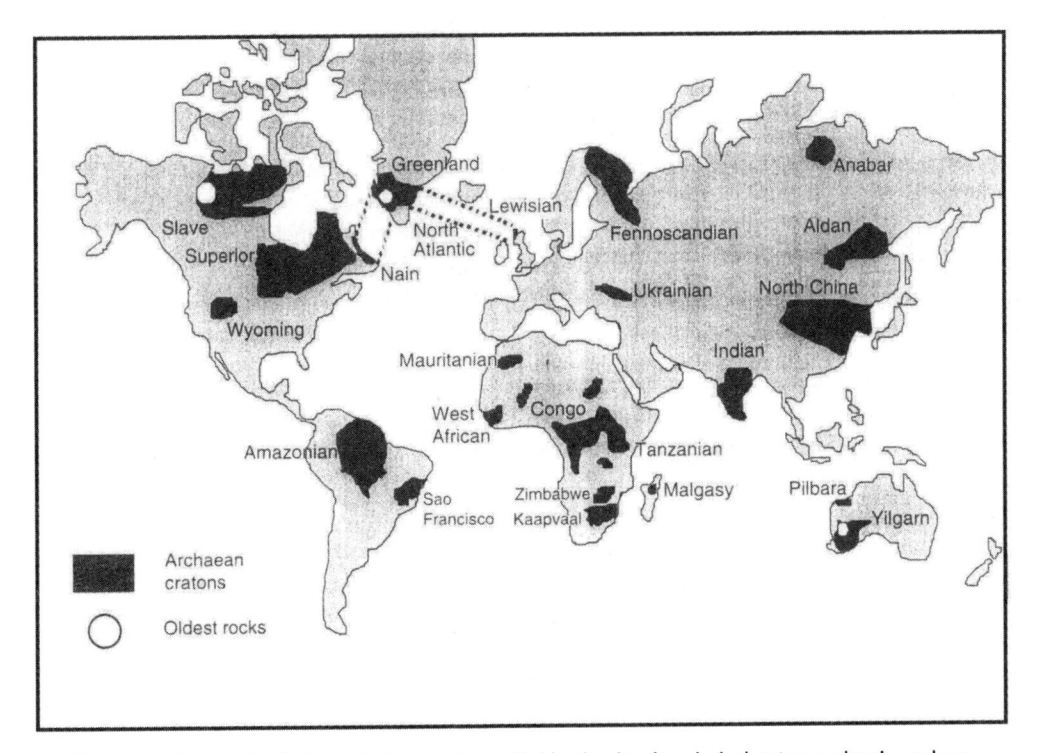

Figure 11.3 The distribution of Archean rocks worldwide, showing the principal cratons and regions where the oldest rocks are preserved. The dashed lines indicate the rocks of the North Atlantic Craton now separated by the Atlantic Ocean (Rollinson, 2007).

In contrast to many Archean rocks, Hadean rocks are extremely rare and are known in outcrop from only a few locations, three of the best studied being:

1. Itsaq Gneiss from West Greenland: These are the first set of rocks to yield ages older than 3.8 Ga and, hence, have attracted much scientific interest. Itsaq gneiss comprise a number of geological units, suchas the Isua greenstone belt, and Amitsaq gneiss. The former contains metabasalts, ultramafic rocks, and clastic and chemical sediments with ages between 3.7 and 3.8 Ga (Nutman et al., 1996). Some mineral grains are even older at 3.86 Ga. The crystalline rocks of Amitsaq gneiss host zircons with ages between 3.78 and 3.81 Ga.

2. The Acasta Gneiss: These rocks are located in the western margin of the Slave Province in northwestern Canada. Bowring et al. (1989) reported zircon with crystallization ages of 3.96 Ga. Later, ion microprobe studies of different parts of a single grain of zircon, from relatively undeformed samples, indicate that the Acasta gneiss contain components that are 4.03 Ga (Bowring and Williams, 1999).

3. Sandstones and conglomerates in the Murchison district of Western Australia: These ancient rocks which were originally deposited as sedimentary rocks about 3 Ga are known as the Mount Narryer and Jack Hills quartzites (Froude et al., 1983). They contain detrital zircons which were transported from some other preexisting source(s) unlike the in-grown zircons in Itsaq or Acasta gneisses. The U-Pb ages of these zircons range from the time of deposition (~3 Ga) of the sedimentary host to values far exceeding it up to 4 Ga with some as old as 4.4 Ga (Compston and Pidgeon, 1986). These 4.4 Ga old zircons are the oldest solid material yet found on the Earth.

Because of their age, Jack Hills zircons have been intensively studied to infer the conditions in which they crystallized. It turns out that the zircons crystallized at a temperature of ~ 700° C, indistinguishable from the temperature at which zircons are known to nucleate and grow in hydrated granitoid magmas. This is powerful evidence for the presence of ample water, if not vast oceans, at the time of the earliest crust of the Earth (Wilde al., 2001). That such ancient zircons have survived to the present speaks of the value of this accessory mineral as versatile time capsule.

A hydrated granitic continental crust dating back to 4.4 Ga, as indicated by Jack Hills zircons, is not likely to be a primary melt from a peridotitic silicate mantle, for such melts would be expected to be basaltic in composition, on the basis of experimental petrology and the early crusts on our planetary neighbours. An important question is whether any crust formed even earlier, presumably from fractional crystallization of the postulated planet-wide magma ocean. This crust would be expected to be basaltic in composition, and also massive enough to extract into it incompatible elements from a large, if not the entire, volume of the of silicate mantle, leaving the latter complementarily depleted in these elements. The differential partitioning of elements into the melt phase will depend on their relative incompatibility. For example, Rb and U are more incompatible than their respective daughter elements, Sr and Pb resulting in higher Rb/Sr and U/Pb ratios in the crust. In contrast, Sm is less incompatible than its daughter, Nd, resulting in a lower Sm/Nd ratio in the crust. The Sr and Pb ratios will, therefore, grow more rapidly, whereas, Nd ratio will grow more slowly in the crust relative to those in the depleted mantle. If a crustal segment and its complementary source mantle portion remained isolated to the present, and can be sampled now for analysis, the time of their mutual separation can be determined from a two-point isochron defined by their present parent and daughter isotope ratios without any knowledge of their common parent. The time of formation of either the crust or the residual mantle can be determined using the reservoir dating principle, only if the initial parent/daughter ratio in the common source is known or assumed, like the assumption of an initially chondritic Hf/W ratio in the Bulk Silicate Earth (BSE). Whereas, the assumption of a chonritic Rb/Sr and U/Pb ratios in the BSE is not valid because of the volatility of Rb and Pb, a chondritic Sm/Nd ratio in the silicate Earth is quite plausible, because Sm and Nd are both lithophile, refractory, adjacent light rare Earth elements with nearly identical outer electronic structures and condensation temperatures.

Of the seven isotopes of Sm, ^{147}Sm and ^{146}Sm are radioactive with useful half-lives. ^{147}Sm alpha (α) decays to ^{143}Nd with a very long half-life of 106 Gy and ^{146}Sm also α decays to ^{142}Nd with a much shorter half-life of 103 My. So, ^{147}Sm is still extant, whereas, ^{146}Sm must have become extinct within the first 500 My of the formation of the Solar System. The growth of radiogenic ^{143}Nd and ^{142}Nd is measured relative to the non-radiogenic isotope, ^{144}Nd. Useful measurement of ^{143}Nd/^{144}Nd requires mass spectrometric precisions of better than 100 ppm, which were realized only in the mid-1970s by Lugmair et al. (1975) and Lugmair and Marti (1977). Even more stringent precisions (better than 10 ppm) are required to resolve natural variations in ^{142}Nd/^{144}Nd. These were achieved only in the early 1990s.

As the short-lived Hf-W and I-Xe systems are best suited to date core formation and mantle outgassing in the early Earth, the ^{146}Sm-^{142}Nd pair is ideal to date any such early crust formation from the BSE. Moreover, the long-lived ^{147}Sm-^{143}Nd system is useful to track the long-term melting history of the Earth. We will consider the former first and defer the latter to the next section.

11.5 ^{142}Nd Evolution in the Earth's Mantle

The very first application of ^{142}Nd isotopes was by Harper and Jacobsen (1992) to the 3.8 Ga Isua sediments, the oldest rock then known. Their report of a small, yet distinct excess of ^{142}Nd/^{144}Nd in these rocks relative to meteorites was initially disputed, but later confirmed (Boyet, 2003; Caro et al., 2003, 2006). The significance of this dramatic result remained unclear until Boyet and Carlson (2005) demonstrated in 2005 that not only old rocks, like the Isua sediments, but most terrestrial rocks of any age have a uniform ^{142}Nd/^{144}Nd ratio that is 20 parts per million higher than in ordinary chondrites, as shown in Figure 11.4. The $\varepsilon(^{142}$Nd) in this diagram is the deviation in parts per 10,000 of ^{142}Nd/^{144}Nd from a terrestrial Nd standard. This implied that the entire mantle, or at least its upper parts sampled by terrestrial rocks, has a uniform excess ^{142}Nd relative to meteorites throughout Earth history. Boyet and Carlson (2005) suggested that the primitive mantle underwent a global melting event to give rise to a basaltic crust with a low Sm/Nd ratio, leaving the remainder of the BSE with a higher Sm/Nd ratio. This crust, presumably formed from a magma ocean, must have occurred long before the extinction of ^{146}Sm, which, along with other Sm isotopes, was inherited by the Earth from the solar nebula. Formation of this crust must have occurred within the first 100 Ma of Earth's beginning for it to produce the observed excess ^{142}Nd. Boyet and Carlson (2005) preferred to call this early crust, enriched in incompatible elements, the Early Enriched Reservoir (EER). As no relict of EER, with a complementary deficit in ^{142}Nd, has yet been found, Boyet and Carlson believe that EER has sunk out of our reach into the deep mantle. Others, such as O'Neill and Palme (2008) agree that a low Sm/Nd crust formed, but it was destroyed on the Earth's surface by impacts in a process they called 'collisional erosion'. Boyet and Carlson also found that lunar basalts have an indistinguishable positive ^{142}Nd anomaly like the Earth's modern mantle. As the moon is believed to have formed from the Earth within 30 My of the Solar System formation, the protocrust must have formed before this time.

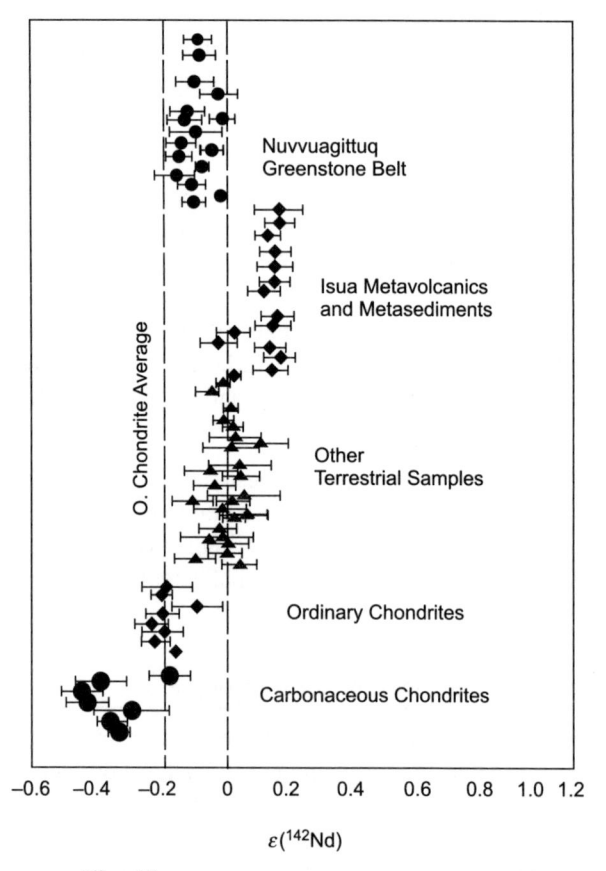

Figure 11.4 Variations of $^{142}Nd/^{144}Nd$ in terrestrial materials and meteorites. All terrestrial rocks younger than 3.5 Ga have uniform $^{142}Nd/^{144}Nd$ ratios that are about 20 ppm (0.2 part in 10,000) higher than in ordinary chondrites (Boyet and Carlson, 2005).

While the remarkable convergence in the timing of the three large scale events—core separation, mantle outgassing, and protocrust generation—provides compelling circumstantial evidence for the former existence of a terrestrial magma ocean, direct or clinching evidence is still lacking.

11.6 ^{143}Nd Evolution in the Earth's Mantle

From Table 5.1 in Chapter 5 the age equation for the decay of ^{147}Sm to ^{143}Nd can be written as:

$$\left(\frac{^{143}Nd}{^{144}Nd}\right) = \left(\frac{^{143}Nd}{^{144}Nd}\right)_t + \left(\frac{^{147}Sm}{^{144}Nd}\right)[\exp(\lambda t) - 1] \quad (11.5)$$

where, λ, the decay constant, is $6.54 \times 10^{-12}\ y^{-1}$ corresponding to a half-life of 106 Gy. The first term on the right side is the Nd ratio, either at any arbitrary time in the past less than the time of formation of the rock or mineral under study. If it represents the Nd isotope ratio at the time of formation, it is conventionally called the initial ratio. ^{146}Sm-^{142}Nd and ^{147}Sm-^{143}Nd could have become another very useful combination of two chemically identical pairs, like the two U-Pb systems covered in Chapter 5, had ^{146}Sm not become extinct in the first 500 My of Earth history. In contrast, ^{147}Sm decays so slowly that only about 3 per cent of its original or primordial abundance has decayed so far.

One can get the initial Nd ratio and t from a Sm-Nd isochron defined by a set of cogenetic samples (rocks or minerals) with a good spread in their parent-daughter ratios. In practice, precise Sm-Nd isochrons are far more difficult to define than Rb-Sr isochrons because of the invariably small spread of parent-daughter ratios in natural objects, compounded by the very small radiogenic ^{143}Nd production due to the slow decay of ^{147}Sm. Precise Sm-Nd isochrons had to wait until Nd ratios could be measured to better than 100 ppm, as mentioned earlier. The reader is referred to the well-known books by Faure and Mensing (2005), Dickin (2005), and Allegre (2008) for many examples of Sm-Nd isochrons. We will restrict ourselves only to large scale Earth processes that can be tracked by the Sm-Nd system.

The Sm-Nd system was first applied to meteorites (Lugmair et al., 1975) in a study which showed that present day ^{147}Sm/^{143}Nd and ^{143}Nd/^{144}Nd ratios in both undifferentiated and differentiated meteorites are quite uniform, with their mean values being 0.1967 and 0.512638, respectively, the latter corrected for mass spectrometric bias using ^{146}Nd/^{144}Nd = 0.7219. These values can be used to calculate the initial Nd ratio from transposing Equation (11.5) as:

$$\left(\frac{^{143}Nd}{^{144}Nd}\right)_t = \left(\frac{^{143}Nd}{^{144}Nd}\right) - \left(\frac{^{147}Sm}{^{144}Nd}\right)[\exp(\lambda t) - 1], \qquad (11.6)$$

and taking t as 4.57×10^9 years. $(^{143}Nd/^{144}Nd)_t$ comes out as 0.506754. It will be recalled from Chapter 5 that this form of the equation is behind the concept of reservoir dating. Time evolution of Nd isotope in a chondritic reservoir is diagrammatically shown in Figure 11.5. According to reservoir concept, any rock derived from this reservoir at anytime would have, at its formation, the same Nd ratio as the reservoir at that time.

The uniformity of parent-daughter and isotope ratios in meteorites, coupled with the strong chemical coherence of Sm and Nd suggested the real and exciting possibility of a chondritic Sm/Nd ratio also in the bulk silicate Earth (primitive mantle) to serve as a well defined initial reference for the Earth. DePaolo and Wasserburg (1976) tested this possibility by using Equation (11.6) to calculate the initial Nd ratios of Archean igneous rocks (like A and B in the figure) of different ages determined independently or from the slope of their Sm-Nd isochrons and plotted them on the diagram. They found that the rocks, with a few exceptions, conformed closely to the CHUR line predicted from meteorites, suggesting that the mantle source of these rocks had a chondritic Sm/Nd ratio. DePaolo and Wasserburg coined the term CHUR, which is the acronym for Chondrite Uniform Reservoir. As a quantitative measure of deviation of sample ratios from CHUR evolution, they introduced the $\varepsilon_t(Nd)$ notation as:

$$\varepsilon_t(Nd) = \frac{\left[\left(\frac{^{143}Nd}{^{144}Nd}\right)_t^s - \left(\frac{^{143}Nd}{^{144}Nd}\right)_t^r\right]}{\left(\frac{^{143}Nd}{^{144}Nd}\right)_t^r} \times 10{,}000 \qquad (11.7)$$

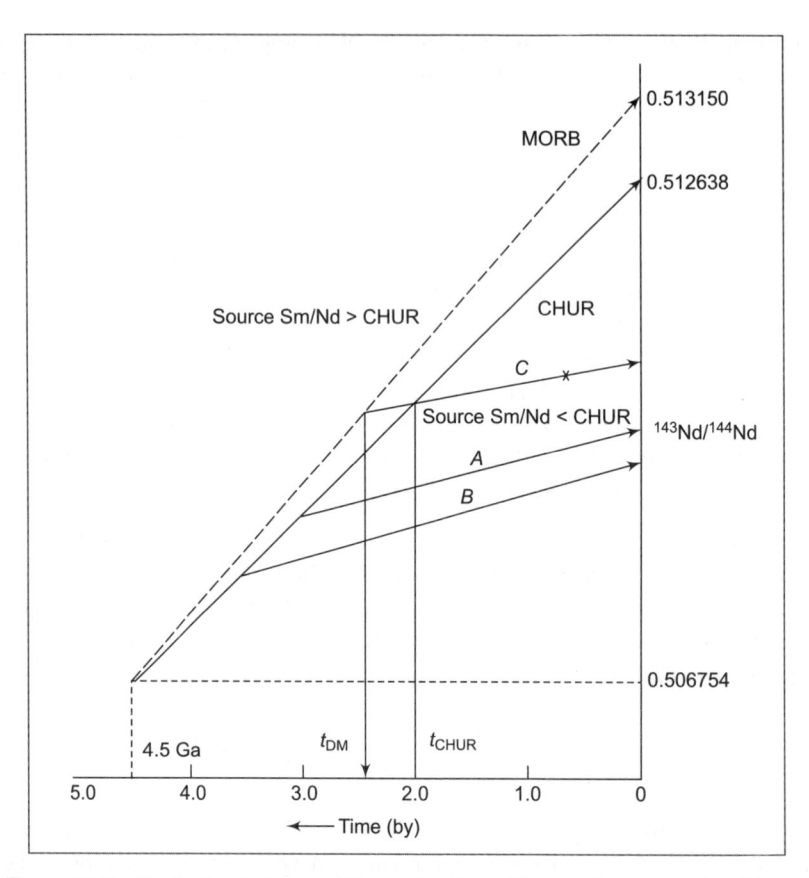

Figure 11.5 Neodymium isotope evolution against time to illustrate the concept of model ages.

where the superscripts s and r refer to sample and CHUR values, respectively, at the same time t in the past. $\varepsilon_t(Nd)$ is simply the relative deviation of the Nd ratio in a rock or mineral sample from the chondritic ratio in parts per 10,000 at the same time in the past. Zero, positive, and negative values of ε_t mean sample derivation from CHUR, a superchondritic reservoir, and a subchondritic reservoir, respectively.

For modern MORBs of essentially zero age, the second term in Equation (11.6) is negligible and, hence, their measured Nd ratios must be exactly those in the mantle source at the present time. DePaolo and Wasseburg (1976) included a few MORB samples in their study, and found that their Nd ratios are not equal to the modern CHUR value at 0.512638, but are distinctly higher, close to 0.51315, corresponding to $\varepsilon_0(Nd) = +10$. This means that at least those parts of the mantle which partially melted to produce MORB were already depleted in their original Nd relative to Sm, because previous episodes of crustal extraction resulted in a net increase of their Sm/Nd ratio. This accounted for the subsequent rapid growth of Nd ratios, as observed. In his later study, DePaolo (1981) found that the Sm/Nd ratio in the mantle began to depart from a strictly chondritic ratio from Proterozoic times (age < 2.5 Ga). He proposed a polynomial of the form $\varepsilon_t(Nd) = 0.25t^2 - 3t + 8.5$ to describe the time evolution of a progressively Depleted Mantle (DM). Goldstein et al. (1984) proposed a linear increase of mantle $^{143}Nd/^{144}Nd$ ratio from 0.506754 at 4.5 Ga to 0.51315 at present as a close enough approximation of the actual evolution of DM. The dotted line above the CHUR trajectory in Figure 11.5 depicts the latter model. Many other DM models have been proposed (Rollinson, 1993).

As the initial Nd ratio of a crustal rock can be calculated from its presently measurable $^{143}Nd/^{144}Nd$ and $^{147}Sm/^{144}Nd$ ratios and its independently determined age when substituted in Equation (11.6), its age can be calculated from the same measured ratios assuming that the rock originated from either a CHUR or DM reservoir. As shown Figure 11.5, the age of a crustal rock like C is the time of intersection of its growth curve with that of either CHUR or the DM. The time since the rock had its Nd ratio on the CHUR line is called its t_{CHUR} model age and the earlier time since its Nd ratio was on the DM curve is called its t_{DM} model age. The general equation given in Chapter 5 for calculating model ages becomes, in the case of Sm-Nd system:

$$t = \left(\frac{1}{\lambda}\right) \ln \frac{\left\{1 + \left[\left(\frac{^{143}Nd}{^{144}Nd}\right)^s - \left(\frac{^{143}Nd}{^{144}Nd}\right)^r\right]\right\}}{\left[\left(\frac{^{147}Sm}{^{144}Nd}\right)^s - \left(\frac{^{147}Sm}{^{144}Nd}\right)^r\right]} \tag{11.8}$$

If a crustal rock has not been preserved as it was originally extracted from the mantle, but was subject to post-formational disturbances like metamorphism and erosion (say) at a time, marked x in Figure. 11.5, the model age of the surviving rock will still refer to the time of separation from the mantle, provided the post-formational disturbances did not alter the parent-daughter ratio of the original rock. As the Sm/Nd ratio is resistant to low temperature metamorphism or even alteration to fine sediment, the Sm-Nd model age of the surviving rock will still give the time its precursor or parent was initially derived from the mantle. For this reason, the Sm-Nd model age is also called the crustal residence age or more explicitly the crustal extraction age. The Rb/Sr and U/Pb ratios are not as resistant to post-formational disturbances. McCulloch and Wasserburg (1978) were the first to undertake a Sm-Nd model age study aimed at measuring the crustal formation ages of several cratonic rock samples, mainly from the Canadian Shield.

Perceptive readers would have noted a significant difference between the ^{143}Nd and ^{142}Nd evidence on when the primitive mantle began to depart from a chondritic Sm/Nd ratio. The ^{143}Nd evidence indicates that the mantle began to depart from an initially chondritic Sm/Nd ratio only gradually with time, whereas, the ^{142}Nd evidence implies a very early and global increase of the mantle Sm/Nd ratio from the primitive mantle. Boyet and Carlson (2005) have discussed how this anomaly can be reconciled.

11.7 ^{87}Sr Evolution in the Earth's Mantle

The age equation for the decay of ^{87}Rb to ^{87}Sr is:

$$\left(\frac{^{87}Sr}{^{86}Sr}\right) = \left(\frac{^{87}Sr}{^{86}S}\right)_t + \left(\frac{^{87}Rb}{^{86}Sr}\right)[\exp(\lambda t) - 1] \tag{11.9}$$

where, λ is the decay constant of ^{87}Rb and $= 1.42 \times 10^{-11}$ y^{-1} (Table 5.1). As Rb is more strongly fractionated from Sr than Sm from Nd during magmatic processes, natural variations in the Rb/Sr ratios between minerals of a single rock or between different rocks solidifying from magma are sufficiently large to yield precise Rb-Sr isochron ages under closed system condition. For this reason, the Rb-Sr method has been the work horse of geochronology of different generations of continental rocks. Readers are referred to well known books for a representative selection from the vast literature of significant Rb-Sr ages of terrestrial rocks. As before, we will restrict ourselves to the application of Rb-Sr system to track mantle evolution.

Although the primitive mantle was unlikely to have a chondritic Rb/Sr ratio, it might appear that the initial Sr ratio of crustal rocks, calculated from their measured (^{87}Sr/^{86}Sr) and (^{87}Rb/^{86}Sr) ratios and ages, could be used to define the Sr evolution curve of the Earth's mantle. But neither the initial Sr ratio, nor the age of a crustal rock, in general, corresponds to its or its precursor's extraction from the mantle. This is because of the ease with which Rb-Sr ages can be reset by post-formational processes. For example, two crustal rocks, granites in particular, can have the same Rb-Sr age, but very different initial Sr ratios. All that can be said with some confidence is that the primitive mantle inherited the initial Sr ratio, 0.6990, of meteorites (Chapter 8), and that the measured Sr ratios of modern oceanic basalts closely match those of their mantle source. The Sr ratio of MORB is 0.70287 ± 0.00045 and that of OIB, 0.70370 ± 0.00089 (White, 2013).

11.8 Coupling Neodymium and Strontium Data

A clever approach to use Sr data is to combine them with Nd data on the same set of rocks at the present time for additional insights into chemical geodynamic processes. Such a coupled study was initiated in the mid-1970s by three scientific teams; one led by G J Wasserburg at the California Institute of Technology, Pasadena, one by R K O'Nions at the Columbia University, New York, and one by C J Allegre at the Institute of Physics of the Globe in Paris.

Distribution of Nd and Sr isotope ratios in different Earth reservoirs is commonly shown as a plot of absolute or relative Nd ratios against absolute Sr ratios. Figure 11.6 is such a plot for rocks of many types and ages (Philpotts, 1990). These include MORB, and OIB from intraplate volcanism in modern upper mantle and deeper mantle, respectively; continental flood basalts(CFB) extruded through old continental crust; solid fragments called xenoliths, or nodules transported by K and other volcanism; very old rocks from ancient cratons; and, finally seawater. Philpotts (1990) brings out the wealth of information from this plot. We will restrict ourselves to a few examples that can be easily understood in the light of the generalized discussion in Chapter 5 on coupling these two chemically different decay pairs.

We first note that the sloping BSE evolution line in this diagram is a special case of the growth curve IJKL in Figure 5.13 for a reservoir with a chondritic Sm/Nd ratio, but an unspecified Rb/Sr ratio. This means that the present Nd ratio in the BSE is 0.512638, but its Sr ratio is unconstrained. However, it must be constrained by the intersection of the BSE line with a horizontal line from the ordinate at 0.512638. In order to locate this intersection, we note that MORB and OIB data define a remarkably linear array, indicating strong inverse correlation between Nd and Sr ratios. Such a negative correlation is, in fact, expected, as partial melting of a mantle segment will result in higher Sm/Nd and lower Rb/Sr ratios in the residual or depleted mantle, and a lower Sm/Nd and higher

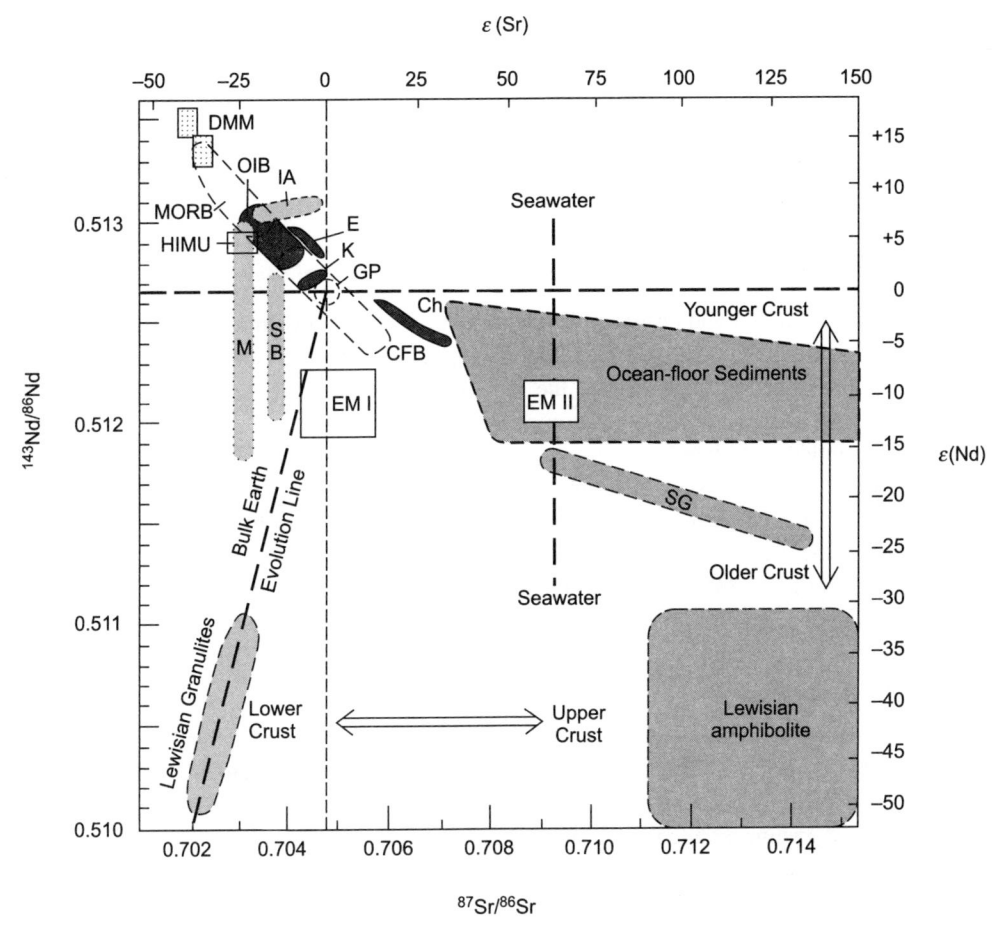

Figure 11.6 $^{143}Nd/^{144}Nd$ vs $^{87}Sr/^{86}Sr$ plot of important rock types. Mid-ocean ridge basalt (MORB), Ocean island basalt (OIB), Island Arc basalts (IA), Kimberlite (K), Garnet peridotite nodule (GP), Continental Flood basalt (CFB), Central volcanic complexes of Mull (M), Skye basalts (SB), Skye granites (SG), Andesites of Equador (E), and Chile (Ch). Possible mantle source components for oceanic magmas shown as rectangles: depleted MORB mantle (DMM), high mu (HIMU), enriched mantle I (EMI), enriched mantle II (EMII) (Philpotts, 1990).

Rb/Sr ratio in the enriched material that is eventually added to the continental crust. Upon subsequent decay of ^{147}Sm to ^{143}Nd and ^{87}Rb to ^{87}Sr over a long enough period, high values of $^{143}Nd/^{144}Nd$ will correlate with low values of $^{87}Sr/^{86}Sr$ in the residual mantle, whereas, there will be low values of $^{143}Nd/^{144}Nd$ and high values of $^{87}Sr/^{86}S$ in the continental crust. The correlation line represents a sampling of mantle isotope ratios, at least from those parts tapped by MORB and OIB magmas. For this reason this range of isotopic values is commonly referred to as a 'mantle array'. The intersection of this mantle array with the horizontal line from the ordinate at 0.512638 gives the present Sr ratio in the BSE as 0.7047. This value can be used in conjunction with 0.6990, taken as the initial ratio for the Earth, to calculate the $^{87}Rb/^{86}Sr$ ratio of the BSE from the equation above as:

$$\left(\frac{^{87}Rb}{^{86}Sr}\right) = \frac{(0.7047 - 0.6990)}{[\exp(1.42 \times 10^{-11} \times 4.5 \times 10^9) - 1]} = 0.075 \qquad (11.10)$$

This corresponds to an elemental Rb/Sr ratio of 0.030, which is much smaller than a typical chondritic Rb/Sr ratio of 0.30. The horizontal and vertical lines in Figure 11.6 passing through the BSE, divide the diagram into four quadrants. Most mantle-derived rocks plot in the upper left quadrant, and most continental crustal rocks plot in the lower right quadrant. The quadrants respectively refer to rocks depleted or enriched in incompatible elements.

Compositions of MORB and OIB samples defining the mantle array indicate variable degrees of depletion of incompatible, or Large Ion Lithophile elements (LIL), in their sources. This variation could occur laterally or vertically within the mantle. The MORBs have formed from the most strongly depleted source. Their restricted compositional range indicates that a rather homogeneous and depleted mantle exists beneath mid-ocean ridges. The OIBs, believed to be plume derived from a deeper mantle region, show a wider range of Sr and Nd values along the mantle array than MORBs. They mostly lie within the region marked OIB in Figure 11.6; but, they can extend to compositions of BSE and even beyond on the continuation of the mantle array. Thus, their source is also depleted as that of MORBs. Their range of composition may result from mixing material derived from depleted and still pristine sources. Because OIBs are believed to develop over mantle plumes, their magmas probably tap a deeper part of the mantle than do MORBs, and, thus, they have more pristine compositions.

The compositions of Continental Flood Basalts (CFB) span a range on either side of BSE. This could be due to: (1) Some CFB magmas were indeed from a pristine mantle; (2) Some magmas originated in a depleted source, but assimilated sufficient enriched continental crust *en route* to the surface to mimic a pristine mantle source; and (3) Some were sourced from a depleted mantle which was enriched by metasomatism (introduction of extraneous material) prior to magmatism. Perhaps all three possibilities were combined in varying degrees in different cases. The nearly pristine compositions of kimberlites (K) and their enclosed garnet peridotitic xenoliths (GP) imply pristine, or least depleted, mantle beneath continental crust where kimberlites originate.

Most crustal rocks plot in the lower right quadrant in Figure 11.6 with a much wider variation of their isotope ratios than in mantle-derived rocks. This is partly due to the differential response of parent-daughter ratios to post-formational surface processes like metamorphism, intra-crustal melting, erosional transport, and sedimentation. In general, the greater the age of a crustal segment, the farther to the lower right of the diagram it plots. Further, Rb/Sr ratios increase upwards in the crust, whereas, the Sm/Nd ratio varies much less with depth. This results in a shift of $^{87}Sr/^{86}Sr$ ratios from right to left with increasing depth in the crust. For example, the deepest-exposed granulites of the 2.95 Ga old Lewsian complex of Scotland have very low Sr ratios (0.702–0.703), whereas, the ambhibolite facies rocks in the upper parts of the complex have high Sr ratios (0.711–0.720).

Simple mixing of mantle-derived magmas with preexisting crustal rocks can be recognized by mixing curves deviating from the direction of the mantle array and skipping the pristine mantle composition. We recall from Chapter 4 that a mixing curve between two end-members of different Nd and Sr isotopic compositions will, in general, be a hyperbola, the sense of its curvature depending on the ratio of Sr/Nd in the two end-members. The mixing curve will be a straight line only when this ratio is about unity. As continental crustal rocks have, in general, lower Sr/Nd ratios than mantle-derived rocks, mixing curves between them will be gently concave upward. As an example, the basalts of the islands of Mull (M) and Skye (SB) on the northwest coast of Scotland erupted during the early opening of the North Atlantic through continental crust of Lewisian metamorphic rocks (referred to above) in contrast to those erupted later from the spreading Mid-Atlantic Ridge. The almost vertical trends of Mull and Skye basalts in Figure 11.6 are interpreted as simple mixing lines between a depleted magma and lower crustal Lewisian granulites. This is a clear case of lower crustal contamination of a later mantle-derived magma.

The granites within the large central igneous complex in Mull and Skye are believed to have formed from the fusion of crustal rocks or contamination of mantle-derived magmas. These granites (SG) plot along a gently curved line between the mantle array and the highly radiogenic Lewisian ambhibolites, and are, therefore, considered to be a mixture of mantle magma and preexisting crustal material. In contrast to the basalts, the granites must have assimilated upper crustal rocks.

Calcalkaline igneous rocks developed over a subduction zone may incorporate one or more of the five possible components, each with distinct isotopic signatures: (1) The mantle wedge overlying the subduction zone; (2) Subducted oceanic crust (MORB); (3) Subducted sediments overlying the oceanic crust; (4) seawater; and (5) crustal material assimilated during their ascent. We will consider the contribution of any of these five components to the isotopic composition of alkaline rocks in different tectonic settings, like island arcs developing between oceanic plates and continental arcs forming between oceanic and continental plates.

The New Britain island arc, far from any continental material, includes rocks ranging from basalt to rhyolite. Their Nd and Sr isotope compositions conform to a linear trend (IA in Figure 11.6) with a slightly positive slope extending from the OIB compositions in the mantle array to the Sr ratio of the BSE. Other island arcs also show this trend of increasing Sr ratios with an almost constant ^{143}Nd/^{144}Nd ratio. A substantial contribution of ocean floor sediments would lead to a trend with a negative slope. The most likely explanation for this isotopic trend is the incorporation of seawater in the island arc magmas. Sea water has a significant Sr concentration (~8 ppm), and its present ^{87}Sr/^{86}Sr ratio is 0.7092 (as shown in the figure). However, the Nd concentration is very low (< 3 × 10^{-5} ppm). Thus, incorporation of sea water in any rock in the mantle array will simply shift its Sr ratio from left to right. Sea water can be incorporated in mid-ocean ridge basalts, and, thus, oceanic crust, directly as pore water, or in its hydrothermally altered minerals. The Nd and Sr isotopes in island arc magmas can then be explained as due to melting of the mantle wedge, fluxed with seawater released from the oceanic plate, as it eventually subducts below the wedge. The slight positive slope of the trend could be due to a small amount of melting of the MORB on the down-going plate.

Sr and Nd isotope patterns of andesites from continental arcs, such as in Equador (E) and Chile (Ch), plot roughly parallel to the mantle array, the former in the depleted quadrant and the latter in the enriched quadrant. This trend can be interpreted as resulting from the mixing of three major components: a depleted mantle source, seawater, and a component that is enriched relative to the BSE composition. This interpretation is not unequivocal because the nature of the enriched component could be oceanic sediments, continental crustal material, or even mantle that was enriched in incompatible elements (Philpotts, 1990). But that the andesites consist predominantly of a mantle derived component is secure.

11.9 ^{206}Pb, ^{207}Pb Evolution in the Earth's Mantle

The evolution of ^{206}Pb and ^{207}Pb from the decay of ^{238}U and ^{235}U, respectively, relative to the nonradiogenic ^{204}Pb, in a chemically closed system is given by the equations:

$$\left(\frac{^{206}Pb}{^{204}Pb}\right) = \left(\frac{^{206}Pb}{^{204}Pb}\right)_t + \left(\frac{^{238}U}{^{204}Pb}\right)[\exp(\lambda_1 t) - 1] \quad (11.11)$$

$$\left(\frac{^{207}Pb}{^{204}Pb}\right) = \left(\frac{^{207}Pb}{^{204}Pb}\right)_t + \left(\frac{^{235}U}{^{204}Pb}\right)[\exp(\lambda_2 t) - 1] \quad (11.12)$$

where, λ_1 and λ_2 are the decay constants of ^{238}U and ^{235}U, respectively. As $\lambda_2 > \lambda_1$, ^{207}Pb/^{204}Pb would have increased more rapidly than (^{206}Pb/^{204}Pb) early in the Earth and much more slowly later. The two curves, defined by these equations, can be combined into one by plotting (^{207}Pb/^{204}Pb) vs (^{206}Pb/^{204}Pb) for various values of t, as shown schematically in Figure 11.7. This is called the growth curve, analogous to the curve IJKL in Figure 5.10. Dividing Equation (11.12) by Equation (11.11), with the first term on the right side transposed to the left side, and remembering that (μ^2/μ^1) is a universal constant (1/137.8 at the present time), we get:

$$\frac{\left(\frac{^{207}Pb}{^{204}Pb}\right) - \left(\frac{^{207}Pb}{^{204}Pb}\right)_t}{\left(\frac{^{206}Pb}{^{204}Pb}\right) - \left(\frac{^{206}Pb}{^{204}Pb}\right)_t} = \left(\frac{1}{137.8}\right)\frac{[\exp(\lambda_2 t) - 1]}{[\exp(\lambda_1 t) - 1]}$$

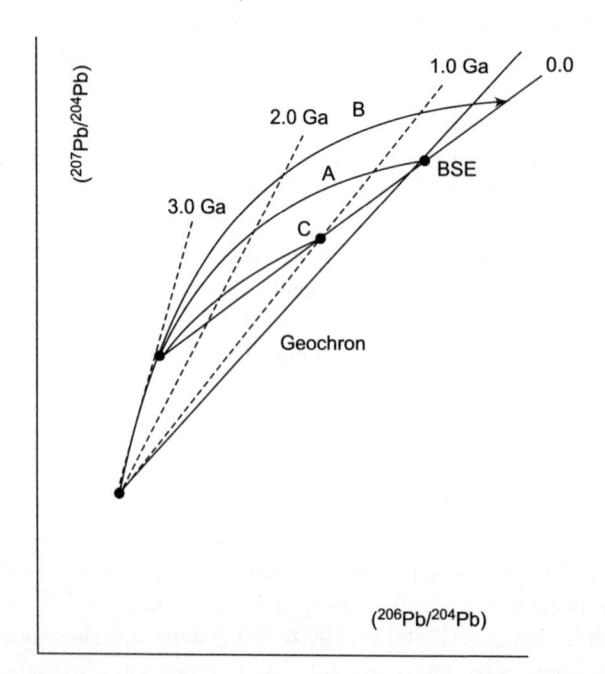

Figure 11.7 Evolution of lead isotope ratios in System A and two of its derivatives, B and C at time t_1.

As the right side is a constant for a given t; this is the equation of a straight line with slope equal to this constant and passing through the point $(^{206}Pb/^{204}Pb)_t, [(^{207}Pb/^{204}Pb)_t]$, as shown in Figure 11.7. If this original chemical system, say A, undergoes a chemical fractionation at a later time into two subsystems B and C with their parent-daughter ratios respectively increased and decreased relative to that of System A, the present-day Pb ratios of B and C will plot away from the above straight line on its either side, but lie along another straight line with a smaller slope, as shown.

Assuming that the Earth evolved at the same time, and from the same materials, as meteorites, and remained closed for 4.55 Gy to the present, its present Pb ratios must plot on the line with a slope corresponding to 4.55 Ga, passing through primordial meteorite ratios (9.307, 10.294 as given in Table 9.1). This line is called the Geochron and is the same for both the Bulk Earth (BE) and BSE, despite their possibly very different U/Pb ratios. The slope of the Geochron will obviously depend on the age assumed for the Earth. The U/Pb ratio in the BSE comes out to be close to 8.5. If a part of the BSE fractionates into two reservoirs, respectively depleted and enriched in incompatible elements, at a later time their present-day Pb ratios will not plot on the Geochron, but on its either side as shown in Figure 11.7. As U is more incompatible than Pb, increases in U/Pb should accompany increases in Rb/Sr and decreases in Sm/Nd and Lu/Hf. In other words, Pb ratios of MORB should plot to the left of the Geochron, and of crustal rocks to its right. Available Pb data on MORB, OIB, and continental crustal rocks are shown in Figure 11.8. The upper crustal rocks do plot, as expected, to the right of the geochron, but, surprisingly, most of the OIBs and many MORBs also plot to the right of the geochron, implying that that the mantle actually experienced an increase and not a decrease in U/Pb ratio. This problem has been termed the Lead Paradox. No satisfactory explanation for this paradox has yet been given (Dickin, 2005). The approximately linear array defined by the oceanic basalts can be interpreted as a Pb-Pb isochron corresponding to an age of about 2.0–1.5 Ga. That, it can as well be a mixing line between the extreme members of this array, without any time significance, cannot be ruled out.

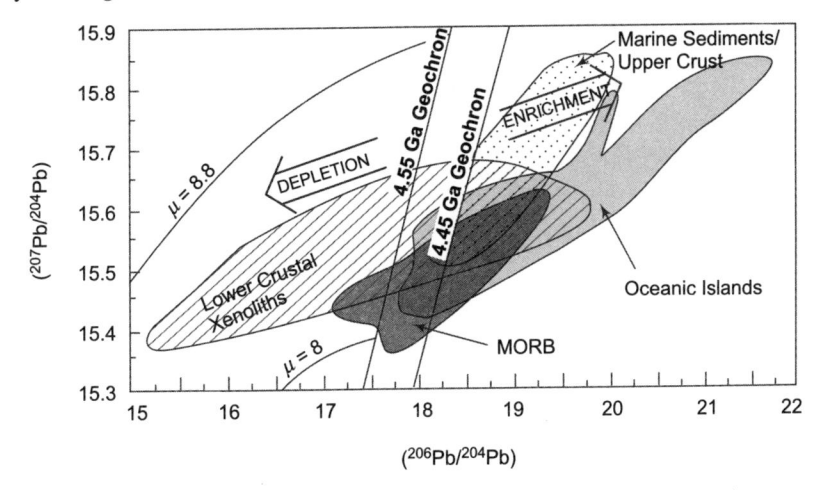

Figure 11.8 Lead isotope ratios in major terrestrial reservoirs (White, 2013).

11.10 Evolution of ^{176}Hf in the Earth's Mantle

The evolution of radiogenic ^{176}Hf relative to the nonradiogenic ^{177}Hf from the decay of ^{176}Lu is given by the equation:

$$\left(\frac{^{176}\text{Hf}}{^{177}\text{Hf}}\right) = \left(\frac{^{176}\text{Hf}}{^{177}\text{Hf}}\right)_t + \left(\frac{^{176}\text{Lu}}{^{177}\text{Hf}}\right) [\exp(\lambda t) - 1] \tag{11.13}$$

where, λ is the decay constant corresponding to a half-life of 35.7 Gy. This half-life is similar to that of ^{87}Rb, but much shorter than 106 Gy for ^{147}Sm. The Lu-Hf system is similar in other respects to the Sm-Nd system because in both cases the elements are relatively immobile and refractory, and the daughter elements are more incompatible than the parent elements during partial melting. Lutetium is a rare earth element, but Hf is not strictly so, and hence the assumption of a strictly chondritic Lu/Hf ratio for the BSE may not be as valid as is the case of the Sm/Nd ratio. Assuming a chondritic Lu/Hf ratio for the BSE, its present ^{176}Hf/^{177}Hf ratio must be close to the chondritic Hf ratio, 0.282786 ± 0.00003. Measurements of Hf ratios became simpler, and quicker, only after the development of the multi-collector inductively-coupled plasma mass spectrometer (MC-ICP-MS) in the last 15 years.

The advantage of the Lu-Hf system over the Sm-Nd system for mantle studies arises from its shorter half-life and the larger fractionation of Lu from Hf than between Sm and Nd during partial melting. This results in the Hf isotope ratio increasing faster than ^{143}Nd/^{144}Nd in the depleted mantle relative to enriched crust. As expected, Hf and Nd ratios define a positive array which passes close to, but not through, a chondritic BSE (White, 2013). The Lu-Hf system has also been used in the same way as the Sm-Nd system to determine the source of magmas above subduction zones.

11.11 Evolution of ^{187}Os in the Earth's Mantle

The evolution of radiogenic ^{187}Os relative to the nonradiogenic ^{188}Os from the decay of ^{187}Re to ^{187}Os given by the equation:

$$\left(\frac{^{187}\text{Os}}{^{188}\text{Os}}\right) = \left(\frac{^{187}\text{Os}}{^{188}\text{Os}}\right)_t + \left(\frac{^{187}\text{Re}}{^{188}\text{Os}}\right) [\exp(\lambda t) - 1] \tag{11.14}$$

where the decay constant λ is 1.55×10^{-11} y^{-1}. The Re-Os system differs from other parent-daughter systems in important ways. First, both Re and Os are siderophile unlike Rb, Sr, Sm, Nd, Lu, Hf, and U, which are lithophile. Thus, the extremely small concentrations of Re and Os in the Earth's mantle relative to chondrites are believed to be due to the almost quantitative extraction of Re and Os of the primitive Earth into the iron core very early on. That the ratio of the residual Re and Os in the mantle is still chondritic in proportion is explained as accretion of the last 1 per cent of material to the Earth after core formation, or leakage of Re and Os from the outer core into the mantle. The second distinguishing property of Re and Os is that Os is compatible and Re is somewhat incompatible, resulting in an enormous difference of the Re/Os ratio from 0.08 in the mantle to ~10 in basalts or crustal rocks. As a consequence, Os isotope ratios range from 0.1275 in the mantle to ~1.2 in crustal materials. The measurement of very small concentrations of Os, at ppb (parts per billion), and ppt (parts per trillion) levels, had to wait until 1981 for development of ultrasensitive rather than ultraprecise techniques (Creaser et al., 1981).

11.12 Evolution of Strontium and Neodymium Isotope Ratios in Seawater

The seawater differs from all other reservoirs in that it is not a product of chemical differentiation of an originally homogeneous silicate mantle (BSE), but a mixture of derivatives from isotopically distinct reservoirs. The prominent and obvious inputs are enriched continental crust and hydrothermal exchanges between seawater and hot basaltic magmas erupting on the seafloor. Figure 11.6 shows that the Sr isotopic composition of seawater at present is extremely homogeneous at 0.7092 through the oceans, but its Nd isotopic composition varies from 0.5110 to 0.5130 across the oceans. This means that Sr is well mixed, but not Nd. The reason for this seemingly impossible scenario lies in the concept of residence time that is well known in elemental geochemistry. The residence time, t_E, of an element, E, in a reservoir is the average time it remains in the reservoir, and is defined by:

$$t_E = \frac{E}{\left(\frac{dE}{dt}\right)} \tag{11.15}$$

where, E is the total amount of E in the reservoir and (dE/dt) is the rate of influx or efflux of the element. If the input and output rates are steady and equal, residence time will be constant over time. The degree of homogeneity of E in the reservoir will depend on the turn-over or mixing time of the reservoir, that is, how fast the old and new material are mixed relative to the residence time. Mixing times much shorter than residence time of an element will ensure its homogeneous distribution in the reservoir, but those comparable to residence time of an element will produce spatial variations in the distribution of that element in the reservoir.

The mixing time of inter-oceanic water masses is about 1,000 years, which is much less than the residence time of Sr of ~ 5 My, but comparable with the residence time of Nd at ~1,000 years. This explains the spatial uniformity of Sr ratios and spatial variability of Nd ratios in seawater. Seawater integrates inputs of Sr from continental runoff, hydrothermal alteration of oceanic basalts, and old sediments and outputs to form new sediments (Veizer, 1989). As these contributions and withdrawals would have varied with time, the Sr isotope ratio of seawater is also expected to vary with time, but remain uniform through the oceans at any particular time. This has been demonstrated from Sr ratio measurements in well-dated organically and inorganically precipitated carbonates. Figure 11.9 shows that the Sr ratio of seawater remained close to mantle values until 2.5 Ga and increased, thereafter (Veizer, 1989). More precise measurements reveal fluctuations in the ratio during the Phanerozoic (< 500 Ma), but a nearly steady increase during the Cretaceous (< 70 Ma) from 0.7077 to the present value of 0.7092. The last part of the seawater curve is now known with high precision and serves as a calibration curve to date unknown carbonate and phosphate samples simply from their measured Sr ratios (Banner, 2004).

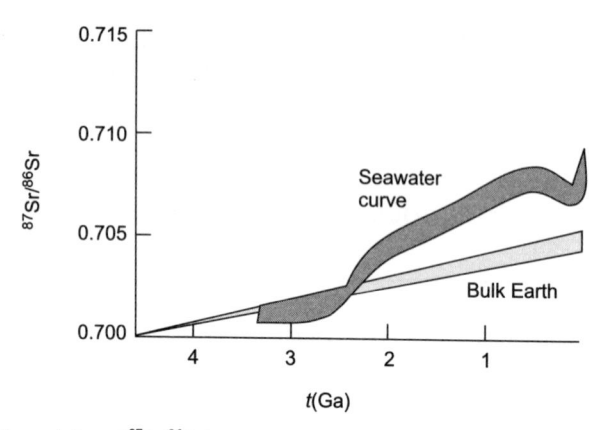

Figure 11.9 The variation of $^{87}Sr/^{86}Sr$ in seawater through time relative to the bulk earth (Veizer, 1989).

The world-wide formation of a thin clay layer marking the Cretaceous-Tertiary (K-T) boundary is now believed to have been caused by a bolide impact on the Earth at this time (Alvarez, 1987). This suggested that the impact, if as massive as proposed, could have left its imprint on the Sr isotope record in ocean sediments. Initial efforts to confirm this gave equivocal results. But Martin and MacDougall (1991) detected a distinct, but small spike on the otherwise smooth long-term variation of Sr ratio in the seawater. The very high contrast in Os isotope ratios between meteorites and crustal rocks, and the similarity of behavior of Sr and Os in, suggest that Os isotope ratios in sea water at the time of bolide impact could provide even stronger evidence for a bolide impact (Peuker Ehrenbrink et al., 1995).

Elements, like Sr, which remain in solution for times of ~10^6 years or longer, will integrate the average input over these long times, thereby smoothing out any regional differences in the Sr concentrations and isotope compositions of inputs, as indeed seen. But elements which are rapidly (< 10^3 years) removed from the water column will show up regional differences, both in concentration and isotopic composition. This is the case, in particular, with Nd (t_E ~1,000 years) and Pb (t_E ~100 years). As Nd concentration in seawater is many orders of magnitude smaller than that of Sr, initial measurements of Nd were made on manganese (Mn) nodules, which secrete Nd from seawater to measurable levels. These, and subsequent direct analyses of seawater, confirmed such regional variations. For example, Piepgras and Wasserburg (1980) measured Nd isotope ratios from 0.511938 to 0.51253 in the Atlantic Ocean and from 0.512442 to 0.51253 in the Pacifc. The lower Nd ratio in the Atlantic reflects Nd supply from felsic rocks of the continental crust ringing the Atlantic basin and the higher Nd ratio in the Pacific supply of REE from predominantly mantle-derived rocks on oceanic islands and on island-arcs in and around the Pacific basin. This inference is strengthened by the Nd composition in Baffin Bay being dominated by drainage from Archean rocks. More detailed studies of Nd isotopes in seawater and river waters are covered in the book by Faure and Mensing (2005).

With the residence time (~100 years) of Pb much shorter than ocean turn-over time, Pb isotope compositions in sea water must reflect even more localized sources of Pb, which has attracted its use as a tracer of environmental changes.

11.13 MAGMA SOURCES IN THE MANTLE

Mid-Ocean-Ridge Basalts are the most abundant volcanics on the Earth produced at ~3 km^3/y in constructive plate margins. They show the highest Nd and lowest Sr ratios which are also remarkably uniform. These features suggest that the MORB source is a major reservoir in the mantle that is very strongly depleted in incompatible elements, situated close to the surface, and well mixed due to efficient convective stirring. In contrast, OIB are much less abundant, show a large dispersion in their Nd and Sr ratios suggesting that their source is inhomogeneous; some parts are depleted, but less than the MORB source, some are nearly pristine, and some parts are actually enriched in incompatible elements relative to the BSE. The variation of OIB isotope ratios between MORB and BSE values can be explained as the result of variable mixing of materials derived from pristine and depleted portions of the mantle. But the presence of enriched components in OIB magmas does not support this possibility. Differences in rare Earth element abundances between the MORB and OIB sources are also likely (White, 2013). Thus, MORB and OIB represent two distinct populations and are derived from two distinct mantle reservoirs. It is also possible that extrusion of MORB was more or less continuous but that of OIB sporadic.

It is clear from the foregoing that the mantle beneath the oceans is isotopically heterogeneous to considerable depths. The present view is that MORB are derived from the shallow mantle, while the OIB are derived from deep mantle and come to the surface by rising plumes. Ocean-Island Basalts are, therefore, particularly important as they can provide insights into the composition of the deep mantle. Zindler and Hart (1986) examined the distribution of Sr, Nd, and Pb data on MORB and OIB samples taken from the literature, making a series of plots like Nd vs Sr, Nd vs Pb, Sr vs Pb, and Pb vs Pb. They then postulated the existence of at least four principal components or magma sources in the mantle, and that the observed variations in isotope ratios result from variable mixing of these components. The four components are: DMM for depleted mantle, EMI for enriched mantle, EMII for enriched mantle II, and HIMU for mantle with a high U/Pb ratio (called μ, by convention). These are shown in three two dimensional plots in Figures 11.10. It is to be noted that no linear array can be recognized in the middle and lower panels. However, the term linear array has survived, and serves now only as a useful reference.

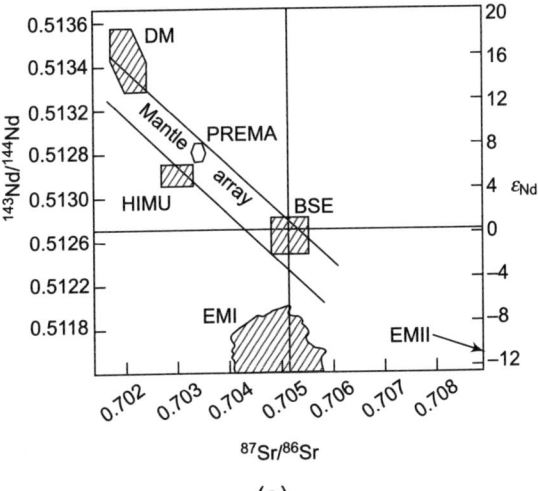

(a)

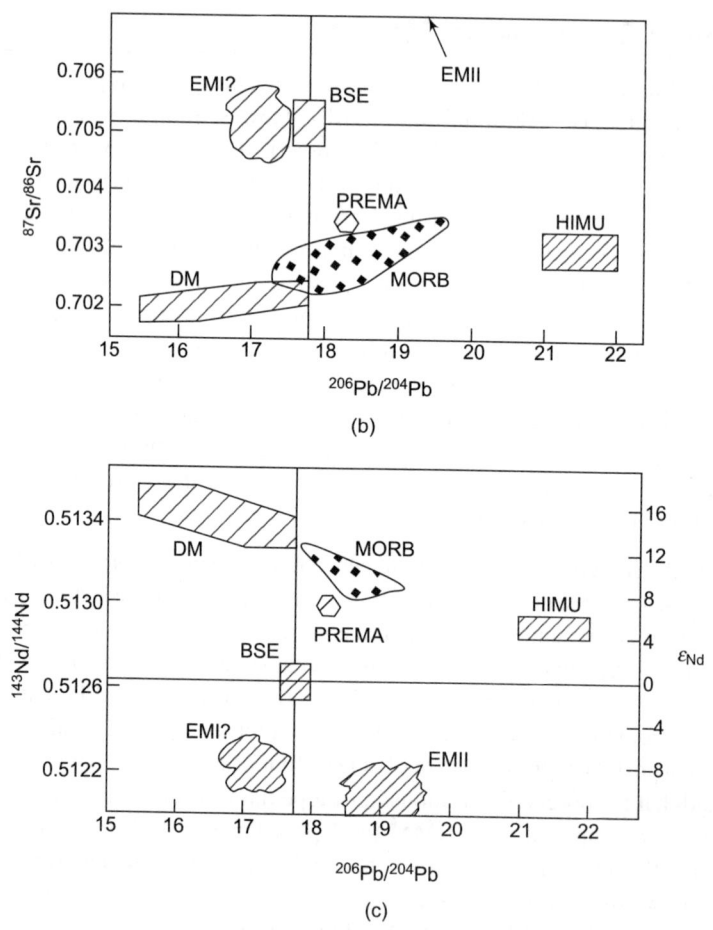

Figure 11.10 Possible mantle source components for oceanic magmas in a plot of (a) ^{143}Nd/^{144}Nd vs ^{87}Sr/^{86}Sr, (b) ^{87}Sr/^{86}Sr vs ^{206}Pb/^{204}Pb, and (c) ^{143}Nd/^{144}Nd vs ^{206}Pb/^{204}Pb. DM, depleted mantle, EMI and EMII, enriched mantle, HIMU, mantle with high U/Pb ratio, BSE, bulk silicate earth and PREMA, frequently observed Prevalent Mantle composition (Zindler and Hart, 1986 from Rollinson, 1991).

Determining the location, size, and evolution of these magma sources is still a major research topic, which is beyond the scope of this book. It is sufficient to note a few possible causes that have been suggested. The MORB mantle, which is the most depleted of all components, is identified with DMM, characterized by high Nd, low Sr, and Pb ratios relative to the BSE. The other three sources show some enrichment in incompatible elements relative to the BSE. In general terms, enrichment is likely to be related to subduction, whereby crustal material is injected into the mantle. HIMU is characterized by high ^{206}Pb/^{204}Pb, but is otherwise similar to DMM. It is thought to be the mantle that acquired a high U/Pb ratio due to recycled (subducted) oceanic crust. Enriched mantle, EMI has high ^{207}Pb/^{204}Pb and ^{208}Pb/^{204}Pb for a given ^{206}Pb/^{204}Pb, low Nd and low Sr ratios. The enrichment in this case could have been caused by metasomatism or recycling of lower continental crust. EMII is similar to EMI, except that its Sr ratio is high, possibly due to recycling of upper continental crust. A fifth, characterized by high ^{3}He/^{4}He ratios, is called Focus zone (FOZO)(not shown in Figure 11.10). This rare gas isotopic signature is found in hotspots, such as Hawaii and Iceland. The primitive ^{3}He is believed to come from a deep mantle source entrained in mantle plumes. Also plotted in Figure 11.10 is another identifiable mantle source christened PREMA (after Prevalant Mantle reservoir) representing abundant data from basalts from ocean islands, intra-oceanic island arcs, and continental basalt suites with Nd ratio = 0.5130, Sr ratio = 0 .7033 and ^{206}Pb/^{204}Pb ratio = 18.2 –18.5. The stippled rectangles in Figure 11.6 represent the Nd and Sr ratios of the components on Figure 11.10. The numerical values of Sr, Nd , Pb, Hf, and Os ratios, as estimated at present, in these sources are given in Table 11.4 (Faure and Mensing, 2005).

Table 11.4 Strontium, Neodymium, Lead, Hafnium and Osmium Isotope Ratios Characteristic of the Four Principal Magma Sources of Mid-Ocean-Ridge Basalts and Ocean-Island Basalts

Ratios	DMM	EM1	EM2	HIMU
$\dfrac{^{87}Sr}{^{86}Sr}$	0.7022	0.7055	0.7075	0.7028
$\dfrac{^{143}Nd}{^{144}Nd}$	0.5133	0.51235	0.51264	0.51285
$\dfrac{^{206}Pb}{^{204}Pb}$	18.2	17.5	19.2	21.7
$\dfrac{^{208}Pb}{^{204}Pb}$	37.7	38.1	39.3	40.7
$\dfrac{^{176}Hf}{^{177}Hf}$	0.28340	0.28265	0.28280	0.28290
$\dfrac{^{187}Os}{^{188}Os}$	0.125	0.152	0.136	0.150

Figure 11.11 shows the preferential distribution of mantle plumes within regions of slow lower mantle seismic velocities. There are two areas where isotopic compositions are particularly extreme (high Sr ratio); one in the south-western Indian Ocean and South Atlantic, and the other in the central Pacific (Hart, 1984; Castillo, 1988). The anomaly in the Indian Ocean is called the DUPAL anomaly, while that in the Pacific is called SOPHITA. Interestingly enough, both anomalies are close to regions where lower mantle velocities are particularly slow. The slow velocity implies that there are regions of high temperatures in the lower mantle. While the exact significance of this remains unclear, it does establish a connection with ocean island volcanism and lower mantle properties, strengthening the plume hypothesis, and favouring a lower origin of plumes (White, 2013).

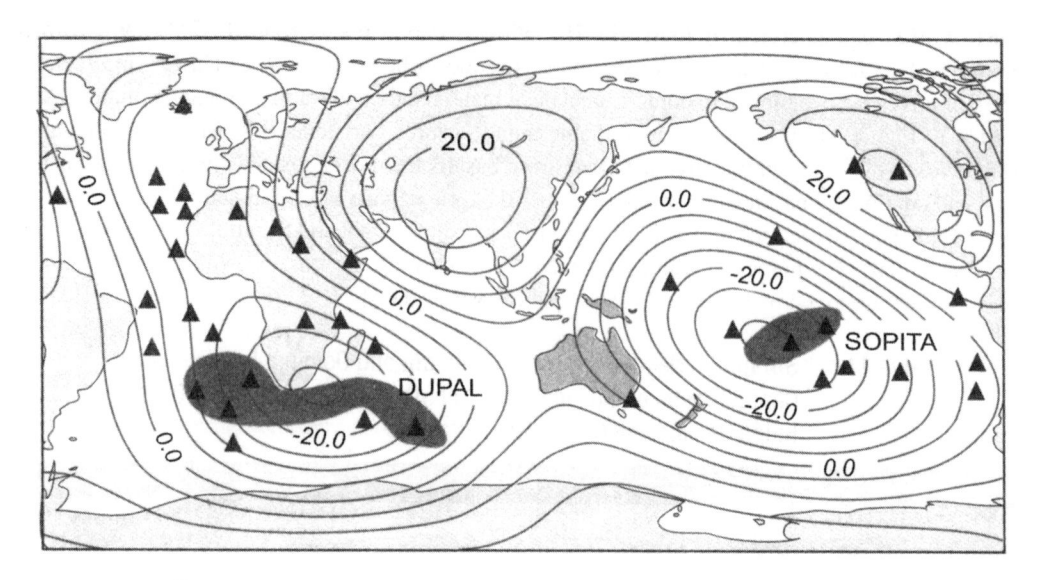

Figure 11.11 Map showing distribution of mantle plumes (triangles) and location of DUPAL and SOPITA isotope anomalies (White, 2013).

Because the DMM is the most strongly depleted source, we can take it as possibly representing that mantle formed by the extraction of the crust. Knowing the volume of the continental crust and its isotopic composition, one can calculate the volume of the mantle with DMM-like composition that would be necessary to provide BSE composition. This shows that as little as about 600 kilometre of the upper mantle need have been involved in the formation of the continental crust; below this it may well be pristine. If the upper mantle is not everywhere as strongly depleted as the DMM, a proportionately greater depth of mantle would be involved, possibly as much as 90 per cent (Allegre et al., 1983).

11.14 Evolution of Radioactive Daughter Isotopes

If the separation of mantle melts and their eruption onto the surface are very recent (< 5,00,000 years), none of the long-lived parent-daughter systems will be useful to date them. Such young magmatic events can, however, be dated by the short-lived daughters of the three long-lived parent-daughter systems, ^{238}U-^{206}Pb, ^{235}U-^{207}Pb, and ^{232}Th-^{208}Pb (Table 5.1). As explained in Chapter 5, they differ from other long-lived systems (^{40}K-^{40}Ar, ^{87}Rb-^{87}Sr, and ^{147}Sm-^{143}Nd) in that the parent isotope in each case does not decay directly to its stable daughter nuclide, but through a number of intermediate radioactive daughter nuclides, with half-lives much shorter than those of their ultimate parents. In an isolated U-Pb system, the two chains of intermediate daughters (Table 5.2) will, in a relatively short time, reach secular equilibrium characterized by equal activity or disintegration rates of each member nuclide, as:

$$\lambda_p p = \lambda_1 d^1 = \lambda_2 d^2 = \lambda_n d^n \tag{11.16}$$

where, $\lambda_p, \lambda_1, \lambda_2 \ldots \lambda_n$ are the decay constants of the parent and daughter nuclides in each chain, and $p, d^1, d^2 \ldots d^n$ are their concentrations, respectively. Once secular equilibrium is established, the rate of disintegration of each parent nuclide will effectively be equal to the rate of production of its stable daughter nuclide. This is the basis of U-Pb and Th-Pb dating, as discussed in Chapter 5.

Subsequent natural geological processes like chemical weathering, sedimentation, partial melting, and crystallization of the resulting magma can disturb the secular equilibrium in a U-Pb system by fractionating the daughter nuclides in each series because of their different chemical affinities. The resulting state of disequilibrium among the intermediate members provides two distinct methods of dating the above geological events, on a time scale ranging from a few years to few million years (Ku, 1976). These are:

1. A member of the decay chain, separated from its immediate parent isotope, will decay subsequently to zero at a rate determined by its half-life; and
2. A member of the decay chain, separated from its immediate daughter, will regenerate the latter to reach secular equilibrium with it at a rate determined by the half-life of the daughter isotope.

Faure and Mensing (2005), and Dickin (2005) provide excellent summaries of many applications of U-series disequilibrium to dating most recent processes on and inside the Earth. We will consider only the application of the decay of ^{230}Th in the decay chain of ^{238}U to dating recently erupted lavas. The ^{238}U decay chain up to ^{230}Th is ^{238}U $\rightarrow$ ^{234}Th $\rightarrow$ ^{234}Pa $\rightarrow$ ^{234}U $\rightarrow$ ^{230}Th $\rightarrow$. As ^{234}Th and ^{234}Pa have extremely short half-lives, and ^{238}U and ^{234}U cannot be fractionated during partial melting, we can treat the system as if ^{238}U decays directly to ^{230}Th. We derived, in Chapter 5, the general expression for the time evolution of a radioactive daughter, d, from its parent, p, as:

$$d = d_t \exp(-\lambda_d t) + [\lambda_p p/(\lambda_d - \lambda_p)][1 - \exp(\lambda_p - \lambda_d)t] \qquad (11.17)$$

where, λ_d is the decay constant of the daughter nuclide and d_t its concentration at time t in the past. For ^{238}U-^{230}Th decay in which $\lambda_{230} \gg \lambda_{238}$, this equation becomes:

$$^{230}\text{Th} = (^{230}\text{Th})_t \exp(-\lambda_{230} t) + \left(\frac{\lambda_{238}\, ^{238}\text{U}}{\lambda_{230}} \right) [1 - \exp(-\lambda_{230})t] \qquad (11.18)$$

The first term on the right side is the residue of partially decayed initial ^{230}Th and the second term the ^{230}Th growth from ^{238}U decay. Expressing $(^{230}$Th$)_t$ explicitly as initial ^{230}Th $(^{230}$Th$)_i$, the above equation becomes:

$$(^{230}\text{Th}) = (^{230}\text{Th})_i \exp(-\lambda_{230} t) + \left(\frac{\lambda_{238}\, ^{238}\text{U}}{\lambda_{230}} \right) [1 - \exp(-\lambda_{230} t)] \qquad (11.19)$$

This is a relation between atomic abundances of ^{230}Th and ^{238}U at time t after the onset of disequilibrium. As even the equilibrium concentration of ^{230}Th is extremely small relative to that of the ^{238}U, which itself is usually a trace isotope in natural objects, this equation is usually written in terms of their comparable activities (within square brackets) by multiplying by λ_{230} throughout as:

$$[^{230}\text{Th}] = [^{230}\text{Th}]_i \exp(-\lambda_{230} t) + [^{238}\text{U}](1 - \exp(-\lambda_{230} t)) \qquad (11.20)$$

This equation between absolute activities can be converted into one between relative activities by dividing throughout by the activity of ^{232}Th [^{232}Th], as it does not decay significantly over the time scale of ^{230}Th activity.

$$\left[\frac{^{230}\text{Th}}{^{232}\text{Th}}\right] = \left[\frac{^{230}\text{Th}}{^{232}\text{Th}}\right]_i \exp(-\lambda_{230}t) + \left[\frac{^{238}\text{U}}{^{232}\text{Th}}\right][1 - \exp(-\lambda_{230}t)]$$

This is the equation for a straight line in a plot of [^{230}Th/^{232}Th] vs (^{238}U/^{232}Th] for constant t (Figure 11.12). The first term is its y-intercept and the second term gives its slope [$1 - \exp(-\lambda_{230}t)$]. Since the slope is proportional to the age, this line is like a conventional isochron. But the y-intercept changes with time, unlike in a conventional isochron, for the following reasons: Minerals crystallizing from a homogeneous magma will in general have different (U/Th) ratios, but the same [^{230}Th/^{232}Th] ratios, defining a horizontal line, as shown in Figure 11.12. Let us consider four such mineral phases a, b, c, and d with a and b having [^{230}Th/^{238}U] activity ratios more than 1, and c and d having this ratio as less than 1. A real or hypothetical point, E, in between these two pairs will have this ratio as unity corresponding to secular equilibrium. As t increases, E, called the equipoint, will remain invariant, whereas, c and d will move vertically upwards and a and b move vertically downwards, as shown. In other words the horizontal isochron will rotate anticlockwise about the point E to become the isochron with a slope of [$1 - \exp(-\lambda_{230}t)$]. As t increases further to more than 3,50,000 years [~ five half-lives of ^{230}Th] the isochron will merge with the line of Slope 1, called the equiline by Allegre and Condomines (1976). The time t given by the intermediate isochron need not necessarily be the time of eruption of this lava as the parent magma could have crystallized before its eruption on the surface. Analytical problems in measuring activity ratios in early work have largely been overcome by measuring isotope abundances directly by thermal ionization mass spectrometry (Chen et al., 1986; Edwards et al., 1987), and more recently by MC-ICPMS (Shen et al., 2002; Stirling et al., 2002).

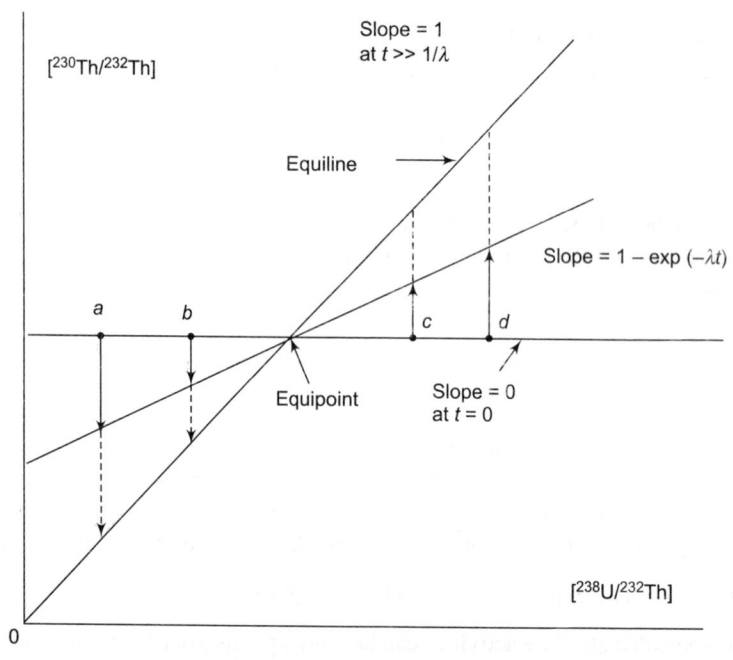

Figure 11.12 Isotopic evolution of igneous rocks on [^{230}Th/^{232}Th] vs [^{238}U/^{232}Th] isochron diagram (Dickin, 2005).

11.15 GIANT IMPACT HYPOTHESIS

The current thinking on planetary formation, as noted in the last chapter, envisages many distinct stages—condensation of a hot gas in the solar nebula into fine dust grains, aggregation of these dust grains through gentle collisions into planetesimals, gravitational agglomeration of planetesimals into few hundred of Mars-sized embryos—all within the first million years of the Solar System history. With the heat generated from accretion and decay of short-lived isotopes, such as ^{26}Al of aluminium (Chapter 9), those embryos would have melted enough for dense metals to segregate from the more buoyant silicates. The few planets, such as observed now, evolved from infrequent and traumatic, but constructive collisions of embryos, a stochastic process that took much longer, perhaps between 10 and 100 million years. In other words, giant impacts between planet-size objects and Mars-sized embryos are essential steps to the final stages of planetary assembly.

It is now widely believed that the moon formed from the debris ejected by the impact of a Mars-sized body with the growing Earth, each with its already segregated iron core. This so-called 'giant impact hypothesis' on the origin of the moon supersedes the previous theories of its origin: capture, fission from the parent Earth, and co-accretion during planetary growth (Dalrymple, 1991). Given the current angular momentum of the earth-moon system, dynamical simulations require that the impact delivered a glancing blow. Most of the projectile's Fe core sank into the Earth's, while its silicate mantle vapourized to blanket the Earth with a rock-vapour atmosphere. This atmosphere cooled and coalesced within about 1,000 years to form the moon (Canup, 2004). The moon must obviously be older than 4,440 Ma, which is the highest Sm-Nd age determined on one of the lunar rocks returned by Apollo astronauts in the 1970s. Dalrymple (1991) has compiled and interpreted the numerous radiometric age results on lunar samples.

We have seen early in this chapter that a difference of about two parts in 10,000 in the W isotope composition between the Earth's mantle and chondrites has been attributed to the segregation of the Earth's core between 30 and 100 million years after the birth of the Solar System, dated at 4,567 Ma (Chapter 9). The upper limit will place the accretion of the earth, therefore, as not later than 4537 Ma. The Hf-W chronometer was later applied to lunar rocks to date the giant, moon-forming-impact. The results revealed virtually no difference in the ^{182}W/^{184}W ratios between the Earth's mantle and the moon, as shown in the Figure 11.2. The data have been modelled to infer that the impact occurred between 30 and 60 million years after the birth of the Solar system, and, hence, consistent with the impact hypothesis. The equality of W ratios provides convincing evidence that the moon inherited its chemical properties from the Earth. Indeed, the mantles of the two bodies are isotopically indistinguishable, at least in those regions that have been sampled so far. That would be no surprise if the bulk of the moon came from the silicate Earth. The snag, however, is that the current simulations of the giant impact hypothesis require that the bulk of lunar material must have come from the Mars-sized impactor rather than the Earth.

The few case histories above are mainly to covey the excitement of research and discovery in modern radiogenic isotope science. Although the number of possible parent-daughter systems is exhausted, there is still a lot of scope for their interesting and imaginative applications by curious and persevering students in many fields.

References

Aitken, M. J. 1985. *Thermoluminescence dating*. London: Academic Press.

Aitken, M. J. 1998. *An Introduction to Optical Dating*. Oxford: Oxford University Press.

Albarede, F. 1998. 'Time-dependent models of U-Th-He and K-Ar evolution and the layering of mantle convection'. *Chem. Geol.* 145: 413-429.

Albarede, F. 2003. *Geochemistry: An Introduction*. Cambridge: Cambridge University Press.

Aldrich, L.T., and G. W. Wetherill. 1958. 'Geochronology by radioactive decay'. *Ann. Rev. Nuclear Sci.* 8 : 257-298.

Allegre, C.J. 1987. 'Isotope geodynamics'. *Earth Planet. Sci. Lett.* 86: 175-203.

Allegre, C.J. 1992. *From Stone to Star*. Massachusetts: Harvard University Press, Cambridge.

Allegre, C.J. 2001. 'Condensed matter astrophysics: constraints and questions on the early development of the solar system'. *Phil. Trans. Roy. Soc.* A359: 2137-2155.

Allegre, C.J. 2008. *Isotope Geology*. Cambridge: Cambridge University Press.

Allegre, C.J, and M. Condomines. 1976. 'Fine chronology of volcanic processes using $^{238}U/^{230}Th$ systematics'. *Earth Planet. Sci. Lett.* 28: 395-340.

Allegre, C.J., T. Staudacher, P. Sarda, and M. Kurz. 1983. 'Constraints on evolution of earth's mantle from rare gas systematics'. *Nature* 303: 762-766.

Allegre, C.J., A. Hofmann, and R. K. O'Nions. 1996. 'The argon constraints on mantle structure'. *Geophys. Res. Lett.* 23: 3555-3557.

Alvarez, L.W. 1987. 'Mass extinctions caused by large bolide impacts'. *Physics Today* 40 : 24-33.

Amelin, Y. 2008. 'U-Pb ages of angrites'. *Geochim. Cosmochim. Acta.* 72: 221-232.

Amelin, Y., J. Connely, and R.E. Zartman. 2009. 'Modern U-Pb chronometry of meteorites'. *Geochim. Cosmochim. Acta* 73: 5212-5223.

Anders, E. 1964. 'Origin, age and composition of meteorites'. *Space Sci. Rev.* 3: 583-714.

Anders, E., and Grevesse, 1989. 'Abundances of the elements; meteoritic and solar'. *Geochim. Cosmochim. Acta* 53: 197-214.

Arnold, J.R., and W.F. Libby. 1949. 'Age determination by content; checks with samples of known age'. *Science* 169 : 227-228.

Atkins, P.W. 1986. *Physical Chemistry*. Oxford: Oxford University Press.

Banner, J.L. 2004. 'Radiogenic isotopes: systematics and applications to earth surface processes and chemical stratigraphy'. *Earth Sci. Revs.* 65: 141-194.

References

Bateman, H. 1910. 'Solution of a system of differential equations occurring in the theory of radioactive transformations'. *Proc. Cambridge Phil. Soc.* 15: 423-427.

Beiser, A. 1973. *Concepts of Modern Physics*. New York: McGraw-Hill.

Bernatowicz, T.J., and R. M. Walker. 1997. 'Ancient stardust in the laboratory'. *Physics Today* 50: 26-32.

Bevington, P.R. 1969. *Data Reduction and Error Analysis for the Physical Sciences*. McGraw-Hill, New York.

Bouvir, A., J. Blichert-Toft, F. Moynier, J. D. Vervoort, and F. Albarede. 2007. 'Pb-Pb dating constraints on the accretion and cooling history of chondrites'. *Gechim. Cosmochim. Acta* 71: 1583-1604.

Bouvir, A., and M. Wadhwa. 2010. 'The age of the solar system redefined by the oldest Pb-Pb age of a meteorite inclusion'. *Nature Geoscience* 3: 637-641.

Bouvir, A., L. J. Williamson, G. A, Brenneka, and M. Wadhwa. 2011. 'New constraints on early solar system chronology from Al-Mg and U-Pb systematics in the unique basaltic achodrite NW2976'. *Geochim. Cosmochim. Acta* 75: 5310-5323.

Bowering, S.A., I. S Williamson, and W. Compston. 1989. '3.96 Ga gneisses from the Slave Province, Northwest Territories, Canada'. *Geology* 17: 977-975.

Bowering, S.A., and I.S. Williams. 1999. 'Priscoan (4.00-4.03 Ga) orthogneiss from northwestern Canada'. *Contib. Mineral. Petrol.* 134: 3-16.

Boyet, M., J. Blicher-Toft, M. Rosing, M. Storey, P. Telouk, and F. Albarede. 2003. '^{142}Nd evidence for early earth differentiation'. *Earth Planet Sci. Lett.* 214: 427-442.

Boyet, M., and R. Carlson. 2005. '^{142}Nd evidence for early (>4.53 Ga) global differentiation of the silicate earth'. *Science* 309: 576-581.

Brazzle, R.H., O. V. Pravdivtseva, A. P. Meshik, and C. M. Hohenberg. 1999. 'Verification and interpretation of the I-Xe chronometer'. *Geochim. Cosmochim. Acta* 63: 739-760.

Brooks. C., S. R. Hart, and I. Wendt. 1972. 'Realistic use of two error regression treatments applied to rubidium-strontium data'. *Rev. Geophys. Space Phys.* 10: 551-577.

Brown, L., J. Klein, R. Middleton, I.S. Sacks, and F. Tera. 1982. '^{10}Be and in island-arc volcanics and implications for subduction'. *Nature* 299: 718-720.

Brown, C.C., and A. Mussett. 1993. *The Inaccesible Earth*. London: Chapman and Hall.

Burbridge, E.M., G.R. Burbridge, W.A. Fowler, and F. Hoyle. 1957. 'Synthesis of elements in stars'. *Rev. Mod. Phys.* 29: 547-647.

Cameron, A.E., D. H. Smith, and R. I. Walker. 1969. 'Mass spectrometry of nanogram size samples of lead'. *Anal. Chem.* 41: 525-526.

Canup, R.M. 2004. 'Origin of terrestrial planets and the earth-moon system'. *Physics Today* 57: 56-63.

Caro, G., B. Bourdon, J.L. Birck, and S. Moorbath. 2003. '^{146}Sm-^{142}Nd evidence from Isua metamorophosed sediments for the early differentiation of the earth's mantle'. *Nature* 423: 428-432.

Caro, G., B. Bourdon, J.L. Birck, and S. Moorbath. 2006. 'High precision ^{142}Nd/^{144}Nd measurements in terrestrial rocks: constraints on the early differentiation of the earth's mantle'. *Geochim. Comochim. Acta* 70: 164-191.

Caro, G., and B. Bourdon. 2010. 'Nonchondrtic Sm/Nd artio in the terrestrial planets; Consequences for the geochemical evolution of the mantle-crust system'. *Geochim. Cosochim. Acta* 74: 3333-3349.

Castillo, P. 1988. 'The Dupal anomaly as a trace of the upwelling lower mantle'. *Nature* 336: 667-670.

Chen, J.H., R.L. Edwards, and G.J. Wasserburg. 1986. '^{238}U, ^{234}U and ^{230}Th in sea water'. *Earth Planet Sci. Lett.* 80: 241-251.

Clayton, R.N., T.K. Mayeda, J.N. Goswamai, and E.J. Olsen. 1991. 'Oxygen isotope studies of ordinary chondrites'. *Geochim. Cosmochim. Acta* 55: 2317-2337.

Clayton, R.N. 2002. 'Self shielding in the solar nebula'. *Nature* 415: 860-861.

Compston, W., and V.M. Oversby. 1969. 'Lead isotopic analysis using a double spike'. *J. Geophys. Res.* 74: 4338-4348.

Compston, W., I. Williams, and C. Meyer. 1984. 'U-Pb geochronology of zircons from lunar breccia 73217 using sensitive high mass-resolution ion microprobe'. *J. Geophys. Res.* 89 (Suppl.): B525–B534.

Compston, W., and R.T. Pidgeon. 1986. 'JackHills, evidence of more very old detrital zircons in western Australia.' *Nature* 321: 766-769.

Condie, K. 2005. *Earth As An Evolving Planetary System.* Amsterdam: Elsevier.

Cowan, A. 1976. 'A natural fission reactor'. *Sci. American* 235: 36–47.

Creaser, R.A., D.A. Papanastassiou, and G.J. Wasserburg. 1991. 'Negative thermal ion mass spectrometry of osmium, rhenium, and iridium'. *Geochim. Cosmochim. Acta* 55: 397–401.

Criss, R.E. 1999. *Principles of Stable Isotope Distribution.* Oxford: Oxford University Press.

Cunningham, M.J. 1981. 'Measurement errors and instrument inaccuracies'. *J. Phys. E. Sci. Instrum.* 14: 901-908.

Dalrymple, G.B. 1991. *The Age Of The Earth.* Stanford: Stanford University Press.

Damon, P.E., D.J. Donahue, B.H. Gore, et al., 1989. 'Radiocarbon dating of the Shroud of Turin'. *Nature* 337: 611-615.

Davies, G.F. 1999. *Dynamic Earth.* Cambridge: Cambridge University Press.

Deming, W.E. 1943. *Statistical Treatment of Data.* New York: John Wiley.

De Laeter, J.R. 1990. 'Mass spectrometry in cosmochemistry'. *Mass Spectrometry Reviews* 9: 453-497.

De Laeter, J.R. 1998. 'Mass spectrometry and geochronology'. *Mass Spectrometry Reviews.* 17: 97-125.

De Wit, M. J., and L.D. Ashwal, 1997. *Greenstone Belts.* Oxford: Oxford University Press.

Dempster, A.J. 1918. 'A new method of positive ray analysis'. *Phys. Rev.* 11: 316-324.

References

DePaolo, D.J., and G. J. Wasserburg. 1976. 'Nd isotopic variations and petrogenetic models'. *Geophys. Res. Lett.* 3: 249-252.

DePaolo, D.J. 1981. 'Neodymium isotopes in the Colorado Front Range and crust-mantle evolution in the Proterozoic'. *Nature* 291: 193-197.

DePaolo, D.J. 1988. *Neodymium Isotope Geochemistry*. Berlin: Springer Verlag.

Dickin, A.P. 2005. *Radiogenic Isotope Geology*, Cambridge: Cambridge University Press.

Dodson, M.H. 1973. 'Closure temperature in cooling geochronological and petrological systems'. *Contrib. Mineral. Petrol.* 40: 259-274.

Duckworth, H.E., R.C. Barber, and V.S. Venkatasubramanian. 1986. *Mass Spectroscopy*. Cambridge: Cambridge University Press.

Dziewonski, A.M, and D.L. Anderson, 1981. 'Preliminary Reference Earth Model'. *Phys. Earth Planet. Interior.* 25: 297-356.

Edwards, R.L., J.H. Chen, and G.J. Wasserburg. 1987. '^{238}U-^{234}U-^{230}Th-^{232}Th systematics and the precise measurement of time over the past 500,000 years'. *Earth Planet. Sci. Lett.* 81: 175-192.

Elmore, D., and F.M. Phillips. 1984. 'Accelerator mass spectrometry for measurement of long-lived isotopes'. *Science* 236: 543-550.

Fairbairn, H.W., P.M. Hurley, and W.H. Pinson. 1961. 'The relation of discordant Rb-Sr mineral and rock ages in an igneous rock to its time of subsequent ^{87}Sr/^{86}Sr metamorphism'. *Geochim. Cosmochim. Acta* 23: 135-144.

Faure, G., and T.M. Mensing. 2005. *Isotopes: Principles and Applications*. New Jersey: John Wiley.

Finkel, R.C., and M. Suter. 1993. 'AMS in the earth sciences'. *Advances Anal. Geochem.* 1: 1-114.

Fleisher, R.L., P.B. Price, and R.M. Walker. 1975. *Nuclear Tracks in Solids*. Los Angeles: University of California Press.

Friedlander, G., and J.W. Kennedy. 1955. *Nuclear and Radiochemistry*. New York: John Wiley.

Froude, D.O., T.R. Ireland, P.D. Kinny, I.S. Williams, W. Compston, I.R. Williams, and J.S. Myers. 1983. 'Ion microprobe identification of 4, 100-4, 200 Myr old terrestrial zircons'. *Nature* 304: 616-618.

Fowler, W.A. 1972. 'New observations and old nucleocosmochronologies'. *Cosmology, Fusion, and Other Matters* (Ed) F. Reines. London: Adam Hilger.

Fowler, C.M.R. 1990. *The Solid Earth*. Cambridge: Cambridge University Press.

Gast, P.W. 1960. 'Limitations on the composition of the upper mantle'. *J. Geophys. Res.* 65: 1287-1297.

Geyh, M.A, and H. Schleicher. 1990. *Absolute Age Determination*. Springer-Verlag, Berlin.

Goldstein, S.L., R.K. O'Nions, and P.J. Hamilton. 1984. 'A Sm-Nd isotopic study of atmospheric dusts and particulates from major river systems'. *Earth Planet. Sci. Lett.* 70: 221-236.

Gopalan, K., and M.N. Rao. 1975. 'Uranium-xenon dating by thermal neutron irradiation'. *Can. J. Earth Sci.* 12: 887-889.

Gopalan, K. 2015. 'Natural radioactivity and geochronometry: An overview'. *J. Appl. Geochem.* 17: 266-289.

Gopel, C., G. Manhes, C.J. Allegre. 1994. 'U-Pb systematics of phosphates from equilibrated ordinary chondrites'. *Earth Planet. Sci. Lett.* 121: 153-171.

Gosset, W. 1908. 'The probable error of a mean'. *Biometrika* 6: 1-25.

Grossman, L. L. 1972. 'Condensation in the primitive solar nebula'. *Geochim. Cosmochim. Acta.* 36: 597-619.

Grun, R. 1989. 'Electron spin resonance (ESR) dating'. *Quat. Int.* 1: 65-109.

Habfast, K. 1983. 'Fractionation in the thermal ion source'. *Int. J. Mass Spectrom. Ion Phys.* 51: 165-189.

Halliday, A.N., J.N. Christensen, D. Lee, C.M. Hall, X. Luo, and M. Rehkamper. 'Multiple-Collector Inductively Coupled Plasma Mass Spectrometry'. *Inorganic Mass Spectrometry*. (Eds.) C.M. Barschick, D.C. Duckworth, and D.H. Smith. New York: Marcel Dekker.

Halliday, A.N., and D. Poricelli. 2001. In search of last planets – the paleocosmochemistry of the inner solar system'. *Earth Planet. Sci. Lett.*192: 545-559.

Harmer, R.E., and B.M. Eglington. 1990. 'A review of the statistical principles of geochronometry: towards a more consistent approach for reporting geochronological data'. *S. Afr. J. Geo.* 93: 845-856.

Harper, C.L., and S.B. Jacobsen. 1992. 'Evidence from coupeld ^{143}Nd-^{147}Sm and ^{142}Nd-^{146}Sm systematics for very early (4.5 Ga) differentiation of the earth's mantle'. *Nature* 360: 728-732.

Hart, S.R. 1984. 'A large scale isotope anomaly in the southern hemisphere mantle'. *Nature* 309: 753-757.

Hart, S.R., and A. Zindler. 1986. 'In search of a bulk earth composition'. *Chem. Geol.* 57: 247-267.

Hart, S.R., and A. Zindler. 1989. 'Isotope fractionation laws: A test using calcium'. *Int. J. Mass Spectrom. Ion Processes* 89: 287-301.

Hess, P.C. 1989. *Origins of Igneous Rocks*. Massachusetts: Harvard University Press.

Heumann, K.G. 1992. 'Isotope dilution mass spectrometry (IDMS) of the elements'. *Mass. Spec. Revs.* 11: 41-67.

Heumann, K.G., S. Eisenhut, S. Gallus, E.H. Hebeda, R. Nusko, A. Vengosh, and J. Walczyk. 1995. 'Recent developments in thermal ionization mass spectrometric techniques for isotope analysis'. *Analyst* 120: 1291-1299.

Heymann, D. 1967. 'On the origin of hypersthene chondrites: ages and shock effects in black chondrites'. *Icarus* 6: 189-221.

Hofmann, A. 1997. 'Mantle geochemistry: the message from oceanic volcanism'. *Nature* 385: 219-229.

Hohenberg, C.M., F. Podosek, and J.H. Reynolds. 1967. 'Xenon-iodine: sharp isochronism in chondrites'. *Science* 156: 233-236.

Holmes, A. 1946. 'An estimate of the age of the earth'. *Nature* 157: 680-684.

References

Houtermans, F.G. 1946. 'Die isotopen-Haufigkeiten im naturlich Blei und das alter de Urans'. *Naturwissenschaften* 32: 185-187.

Ingram M.G., and W.A. Chupka. 1953. 'Surface ionization source using multiple filaments'. *Rev. Sci. Instrum.* 24: 518-520.

Jeffery, P.M, and J.H. Reynolds. 1961. 'Origin of excess ^{129}Xe in stone meteorites'. *J. Geophys. Res.* 66: 3582-3583.

Jull, A. J. T. 2001. 'Terrestrial ages of meteorites. In: B. Peuker-Ehrenbrink and B. Schmitz (eds.)' *Accretion of Extraterrestrial Matter Throughout Earth's History,* 241-266. New York: Kluwer Academic/Plenum Publishers.

Kaplan, I. 1955. *Nuclear Physics.* Massachusetts: Addison-Wesley.

Kleine, T., C. Munker, K. Mezger, and H. Palme. 2002. 'Rapid accretion and early core formation on asteroids and the terrestrial planets from Hf-W chronometry'. *Nature* 418: 952-954.

Ku, H.H. 1969. 'Statistical concepts in metrology'. *National Bureau of Standards Special Publication* 300, 1: 296-330.

Ku, T.L. 1976. 'The uranium series methods of age determination'. *Ann. Rev. Earth Planet. Sci.* 4: 347-379.

Kuroda, P.K. 1956. 'On the nuclear physical stability of the uranium minerals'. *J. Chem. Phys.* 25: 981-982.

Lal, D., and B. Peters, 1967. 'Cosmic-ray produced radioactivity on the earth'. In *Handbook of Physics* 16: 551-612.

Lal, D. 1988. 'In situ-produced cosmogenic isotopes in terrestrial rocks'. *Ann. Rev. Earth Planet. Sci.* 16: 355-388.

Langmuir, C.H., R.D. Vocke, G.N. Hanson, and S.R. Hart. 1977. 'A general mixing equation: application to the petrogenesis of basalts from Iceland and Reykjanes Ridge'. *Earth Planet. Sci. Lett.* 37: 380-392.

Langmuir, C.H., and Broecker, W. 2012. *How to Build a Habitable Planet.* Princeton: Princeton University Press.

Lee, Y., Papanastassiou, D., and Wasserburg, G.J. 1977. 'Aluminuium-26 in the early solar system: fossil or fuel?' *Astrophys. J. Lett.* 211: L107-110.

Lee, D.C., and A.N. Halliday. 1995. 'Hafnium-Tungsten chronometry and the timing of terrestrial core formation'. *Nature* 378: 771-774.

Leighton, R.B. 1959. *Principles of Modern Physics.* New York: McGraw-Hill.

Lewis, J.S. 2004. *Physics and Chemistry of the Solar System.* Oxford: Elsevier.

Lineweaver, C.H. 1999. 'A younger age for the Universe'. *Science* 284: 1503-1507.

Ludwig, K.R. 2000. *Isoplot/Ex Version 2.4. A geochronological tool kit for Microsoft Excel*, Sp. Publication 56. Berkeley: Berkeley Geochronolgical Centre.

Lovelock, J.E. 1979. *Gaia. A New Look at Life on Earth.* Oxford: Oxford University Press.

Lovelock, J.E. 1988. *The Ages Of Gaia: A Biography Of Our Living Earth.* Oxford: Oxford University Press.

Lugmair, G.W., N.B. Scheinin, and K. Marti. 1975. 'Search for extinct ^{146}Sm. Pt.I The isotopic abundance of ^{143}Nd in the Juvinas meteorite'. *Earth Planet. Sci. Lett.* 27: 9479-9484.

Lugmair, G.W., and K. Marti, 1977. 'Sm-Nd-Pu time-pieces in the Angra dos Reis meteorite'. *Earth Planet. Sci. Lett.* 35: 273-284.

Lupton, J.G., and H. Craig. 1975. 'Excess ^{3}He in oceanic basalts: evidence for terrestrial primordial helium'. *Earth. Planet. Sci. Lett.* 26: 133-139.

Mahan, B.H. 1975. *University Chemistry*. Massachusetts: Addison-Wesley.

Marshall, B.D., and D.J. DePaolo. 1982. 'Precise age determination and petrogenetic studies using the K-Ca method'. *Geochim. Cosmochim. Acta* 46: 2537-2545.

Martin, E.E,, and J.D. Macdougall. 1991. 'Seawater Sr isotopes at the Cretaceous–Tertiary boundary'. *Earth. Planet. Sci. Lett.* 104: 166-180.

McCulloch, M.T., and G.J. Wasserburg. 1978. 'Sm-Nd and Rb-Sr chronology of continental crust formation'. *Science* 200: 1002-1011.

McDougall, I., and M. Honda. 'Primordial noble gas component in the earth'. *The Earth's Mantle: Composition, Structure and Evolution*, Ed. Jackson, I.N.S. Cambridge: Cambridge University Press.

McDougall, I., and T.M. Harrison. 1999. *Geochronology and Thermochronology by the $^{40}Ar/^{39}Ar$ Method*. Oxford: Oxford University Press.

McIntyre, G.A., C. Brooks, W. Compston, and A. Turek. 1966. 'The statistical assessment of Rb-Sr isochrons.' *J. Geophys. Res.* 71: 5459-5468.

McKeegan, K.,D., M. Chaussidon, and F. Robert. 2000. 'Incorporation of short-lived ^{10}Be in calcium-aluminium rich inclusion from the Allende meteorite'. *Science* 289: 1334-1337.

McSween, H.Y., and G. Huss. 2010. *Cosmochemistry*. Cambridge: Cambridge University Press.

Merrihue, C., and G. Turner. 1966. 'Potassium-argon dating by activation with fast neutrons'. *J. Geophys. Res.* 71: 2852-2857.

Meshik, A. 2005. 'The workings of an ancient nuclear reactor'. *Sci. American* 293: 57-63.

Minster, J.F., J.L. Birck. and C.J. Allegre. 1982. 'Absolute age of formation of chondrites studied by the ^{87}Rb-^{87}Sr method'. *Nature* 300: 414-419.

Mitchell, J.G. 1983. 'Krypton isotopes in neutron irradiated minerals: Use in evaluating sample for ^{87}Rb/^{86}Sr dating'. *Geof. Int.* 22: 137-144.

Nicolaysen, L.O. 1961. 'Graphic interpretation of discordant age measurements on metamporphic rocks'. *Ann. N.Y. Acad. Sci.* 91: 198-206.

Nier, A.O. 1938. 'Variations in the relative abundances of the isotopes of common lead from various sources'. *J. Am. Chem. Soc.* 60: 1571-1576.

Nier, A.O. 1939. 'The isotopic constitution of radiogenic leads and the measurement of geologic time'. Pt.II. *Phys. Rev.* 55: 153-163.

Nier, A.O. 1940. 'A mass spectrometer for routine isotope abundance measurements'. *Rev. Sci. Instrum.* 11: 212-16.

References

Nier, A.O. 1947. 'A mass spectrometer for isotope and gas analysis'. *Rev. Sci. Instrum.* 18: 398-411.

Nier, A.O., E.P. Ney, and M.G. Inghram. 1947. 'A null method for the comparison of two ion currents in a mass spectrometer'. *Rev. Sci. Instrum.* 18: 294-297.

Nutman, A.P., V.R. McGregor, C.R.L. Friend, V.C. Bennett, and P.D. Kinny. 1996. 'The Itsaq Gneiss complex of southern west Greenland: the world's most extensive record of early crustal evolution (3900-3600 Ma)'. *Precambrian Res.* 78: 1-39.

O'Neill, H.St.C, and H. Palme. 2008. 'Collisional erosion and the nonchondritic composition of the terrestrial planets'. *Phil. Trans. R. Soc.* A366: 4205-4238.

Ozima, M., and K. Zahnle. 1993. 'Mantle degassing and atmospheric evolution noble gas view'. *Geochemical J.* 27: 185-200.

Ozima, M., and F.A. Podosek. 1983. *Noble Gas Geochemistry*. Cambridge: Cambridge University Press.

Ozima, M. 1994. 'Noble gas state in the mantle'. *Rev. Geophys.* 32: 405-426.

Papanastassiou, D.A., G.W. Wasserburg, and D.S. Burnett. 1969. 'Initial strontium isotopic abundances and the resolution of small time differences in the formation of planetary objects'. *Earth Planet. Sci. Lett.* 5: 361-376.

Patchett, P.J. 1983. 'Importance of the Lu-Hf system in studies of planetary chronology and chemical evolution'. *Geochim. Cosmochim.* Acta 47: 81-91.

Patterson, C.C. 1956. 'Age of meteorites and the earth'. *Geochim. Cosmochim. Acta* 10: 230-237.

Pauling, L. 1988. *General Chemistry*. New York: Dover Publications.

Penzias, A.A., and R.W. Wilson. 1965. 'A measurement of excess antenna temperature at 4080 Mc/s'. *Astrophys. J.* 142: 419-421.

Percival, J., and G.F. West. 1994. 'The Kapuskasing uplift: a geological and geophysical synthesis.' *Can. J. Earth Sci.* 31: 1256-1286.

Peuker Ehrenbrink, B., G. Ravizza, and A.W. Hoffmann. 1995. 'The marine $^{187}Os/^{188}Os$ record of the past 80 million years.' *Earth Planet. Sci. Lett.* 130: 155-167.

Philpotts, A.R. 1990. *Principles of igneous and metamorphic petrology*. New Jersey: Prentice Hall.

Piepgras, D.J., and G.J. Waserburg. 1980. 'Neodymium isotopic variations in seawater'. *Earth Planet. Sci. Lett.* 50: 128-138.

Potts, P.J. 1987. *A Handbook Of Silicate Rock Analysis*. Glasgow: Blackie.

Pretorius, D.A. 1973. 'The role of EGRU in mineral exploration in South Africa'. *Economic Geology Unit, University of Witwatersrand Information* Circular. 77: 16.

Price, B., and R.M. Walker. 1962. 'Observation of fossil particle tracks in natural micas'. *Nature* 196: 732-734.

Price, P.B., and R. Walker. 1962. 'Chemical etching of charged particle tracks in solids'. *J. Appl. Phys.* 33: 3407-3409.

Ragland, P.C. 1989. *Basic Analytical Petrology*. Oxford: Oxford University Press.

Rehkamper, M., Schonbachler, and H. Stirling. 2000. 'Multiple Collector ICP-MS: Introduction to instrumentation, measurement techniques and analytical capabilities'. *Geostandards Newsletter* 25: 23-40.

Reynolds, J.H. 1960. 'Determination of the age of the elements'. *Phys. Rev. Lett.* 4: 8-10.

Richardson, S.M., and H.Y. McSween. 1989. *Geochemistry: Pathways and Processes*. New Jersey: Prentice Hall.

Rollinson, H.R. 1993. *Using Geochemical Data*. Harlow: Longman.

Rollinson, H.R. 2007. *Early Earth Systems*. Massachusetts: Blackwell Publishers.

Ross, J.E., and L.H. Aller. 1976. 'The chemical composition of the Sun'. *Science* 191: 1223-1229.

Rutherford, E. 1906. *Radioactive Transformations*. New York: Charles Scribner Sons.

Rutherford, E., and F. Soddy. 1902a. 'The cause and nature of radioactivity'. Pt.1 *Phil. Mag.* 4: 370-396.

Rutherford, E., and F. Soddy. 1902b. 'The cause and nature of radioactivity'. Pt.1 *Phil. Mag.* 4: 569-585.

Rutherford, E. 1919. 'Collision of alpha particles with light atoms, IV. An anomalous effect in nitrogen'. *Phil. Mag.* 3: 581-587.

Rutherford, E. 1929. 'Origin of actinides and age of the earth'. *Nature* 123: 313-314.

Rutherford, E., J. Chadwick, and C.D. Ellis. 1930. *Radiations From Radioactive Substances*. Cambridge: Cambridge University Press.

Schmitz, B., M. Tassinari, and B. Peucker-Ehrenbrink. 2001. 'A rain of ordinary chondritic meteorites in the Early Ordovician'. *Earth Planet. Sci. Lett.* 194: 1-15.

Schoenberg, R., B. Kamber, D. Collerson, and O. Eugster. 2002. 'New W–isotope evidence for rapid terrestrial accretion and very early core formation'. *Geochim. Cosmochim. Acta* 66: 3151-3160.

Schramm, D.N., and G.W. Wasserburg. 1970. 'Nucleochronologies and the mean age of the elements'. *Astrophys. J.* 162: 57-69.

Schramm, D.N. 1974a. 'Nucleo-cosmochronolgy'. *Ann. Rev. Astron. Astrophys.* 12: 303-406.

Schramm, D.N. 1974b. 'The age of the elements'. *Sci. American* 230: 69-77.

Schumm, S.A. 1991. *To Interpret the Earth*. Cambridge: Cambridge University Press.

Sharp, Z. 2007. *Principles of Stable Isotope Geochemistry*. Upper saddle River: Prentice Hall.

Shen, C.C., R.L. Edwards, H. Cheng, J.A. Dorale, R.B. Thomas, S.B. Moran, S.E. Weinstein, and H.N. Edmonds. 2002. 'Uranium and thorium isotopic and concentration measurements by magnetic sector inductively-coupled plasma mass spectrometry'. *Chem. Geol.* 185: 165-178.

Shu, F.S. 1982. *The Physical Universe: An Introduction to Astronomy*. Mill Valley: University Science Books.

Silk, E.C.H., and R.S. Barnes, 1959. 'Examination of fission fragment tracks with an electron microscope'. *Phil. Mag.* 4: 970-972.

References

Srinivasan, G., A.A. Ulyanov, and J.N. Goswami. 1994. '^{41}Ca in the early solar system'. *Astrophys. J (Lett.).* 431: L67-70.

Stacey, J.S., and J.D. Kramers. 1975. 'Approximation of terrestrial lead isotope evolution by a two-stage model'. *Earth. Planet. Sci. Lett.* 26: 207-221.

Staudigel, H. et al. 1998. 'Geochemical Earth Reference Model (GERM): description of the initiative'. *Chem. Geol.* 145: 153-159.

Steiger, R.H., and E. Jager. 1977. 'Subcommission on Geochonology: Convention on the use of decay constants in geo- and cosmochronology'. *Earth Planet. Sci. Lett.* 36: 359-362.

Stirling, C.H., D.C. Lee, J.N. Christensen, and A.N. Halliday. 2002. 'High precision in situ ^{238}U-^{234}U-^{230}Th isotopic analysis using Laser ablation multiple-collector ICPMS'. *Geochim. Cosmochim. Acta* 64: 3737-3750.

Tatsumoto, M., R.J. Knight, and C.J. Allegre. 1973. 'Time differences in the formation of meteoites as determined from the ratio of lead-207 to lead-206'. *Science* 180: 1279-1283.

Tera, F., and G.J. Wasserburg. 1974. 'U-Th-Pb systematics on lunar rocks and inferences about lunar evolution and the age of the moon'. Proc. 5th Lunar Conf. (Supt.5). *Geochim. Cosmochim. Acta* 2: 1571-1599.

Thiemens, M.H. 2000. 'History and applications of mass independent isotope effects'. *Ann. Rev. Earth Planet. Sci.* 34: 217-262.

Thiemens, M.H., and J.E. Heidenrich. 1983. 'The mass dependent fractionation of oxygen – A novel isotopic effect and its cosmochemical implications'. *Science* 219: 1073-1075.

Thompson, N, ed. 1990. *Thinking Like a Physicist: Physics Problems for Undergraduates.* Bristol: Adam Hilger, IOP Pub. Ltd.

Tilton, G.R. 1960. 'Volume diffusion as mechanism for discordant lead ages'. *J. Geophys. Res.* 65: 2933-2945.

Touboul, M., T. Kleine, R. Boudon, H. Palme, and Wieler, R. 2007. 'Late formation and prolonged differentiation of the moon inferred from W isotopes in lunar metals'. *Nature* 450: 1206-1209.

Touboul, M., T. Kleine, R. Boudon, H. Palme, and R. Wieler. 2009. 'Tungsten isotopes in ferroan anorthosites: Implications for the age of the moon and lifetime of its magma ocean'. *Icarus* 199: 245-249.

Trieloff, M., M. Falter, and E.K. Jesseberger. 2003. 'The distribution of mantle atmospheric argon in oceanic basaltic glasses'. *Geochim. Cosmochim. Acta* 67: 1229-1245.

Van Schmus, W.R., and J.A. Wood. 1967. 'A chemical and petrologic classification for the chondritic meteorites'. *Geochim. Cosmochim. Acta* 31: 747-765.

Veizer, J. 1989. 'Strontium isotopes in seawater through time'. *Ann. Rev. Earth Planet. Sci.* 17: 141-167.

Villa, I.M., and P.R. Renne. 2005. 'Decay constants in geochronology'. *Episodes* 28: 50-51.

Vollmer, R. 1976. 'Rb-Sr and U-Th-Pb systematics of alkaline rocks from Italy'. *Geochim. Cosmochim. Acta* 40: 283-296.

Wadhwa, M., Y. Amelin, and O. Bogdanowski. 2009. 'Ancient relative and absolute ages for a basalic meteorite'. *Geochim. Cosmochim. Acta* 73: 5189-5201.

Wasserburg, G.J. 1962. 'Diffusion processes in lead-uranium systems'. *J. Geophys. Res.* 68: 4823-4846.

Wasserburg, G.J., D. Papanastassiou, E.V. Nenow, and C.A. Bauman. 1969. 'A programmable magnetic field mass spectrometer with online data processing'. *Rev. Sci. Instrum.* 40: 288-295.

Wasserburg, G.J., S.B. Jacobsen, D.J. DePaolo, M.T. McCulloch, and T. Wen. 1981. 'Precise determination of Sm/Nd ratios, Sm and Nd isotopic abundances in standard solutions'. *Geochim. Cosmochim. Acta* 45: 2311-2323.

Wasserburg, G.J. 1987. 'Isotopic abundances: Inferences on solar system and planetary evolution'. *Earth Planet. Sci. Lett.* 86: 204-265.

Wasson, J.T. 1974. *Meteorites*. Berlin: Springer-Verlag.

Wasson, J.T. 1985. *Meteorites: Their Record of Early Solar System History*. New York: W.H. Freeman.

Wetherill, G.W. 1956. 'Discordant uranium-lead ages'. *Trans. Am. Geophys. Union.* 37: 320-327.

Wetherill, G.W. 1968. 'Stone meteorites: time of fall and origin'. 1969. *Science* 159: 79-82.

Wetherill, G.W. 1975. 'Radiometric chronology of the early solar system'. *Ann. Rev. Nucl. Sci.* 25: 283-324.

Wetherill, G.W., C. Allegre, and C. Brooks, P. Eberhardt, S.R. Hart, V. Ramamurthy, F. Tera, and W.R. Van Schmus. 1981. 'Radiogenic and stable isotopes; radiometric chronology, and basaltic volcanism'. *Basaltic volcanism on the Terrestrial Planets*, Ed. Basaltic volcanism study project, Lunar Planetary Institute. New York: Pergamon Press.

Wetherill, G.W. 1990. 'Formation of the Earth'. *Ann. Rev. Earth Planet. Sci.* 18: 205-256.

Wieler, R., T. Graf, 2001, 'Cosmic rayexposure history of meteorites. In B. Peueker-Ehrenbrink and B. Schmitz, eds'. *Accretion of Extraterrestrial Matter throughout the Earth's history*, 21-40. New York: Klewer Academic/Plenum.

Wilde, S.A., J.W. Valley, W.H. Peck, and C.M. Graham. 2001. 'Evidence for detrital zircons for the existence of continental crust and oceans on the earth 4.4 Gyr ago'. *Nature* 409: 175-178.

Williamson, J.H., 1968. 'Least-squares fitting of a straight line'. *Can. J. Phys.* 46: 1845-1847.

White, W.M. 2013. *Geochemistry*. New York: John Wiley.

Yin, Q., S.B. Jacobsen, K. Tamashita, J. Blichert-Toft, P. Telouk, and F. Albarede. 2002. 'A short time scale for terrestrial planet formation from Hf-W chronometry of meteorites'. *Nature* 418: 949-952.

York, D. 1966. 'Least squares fitting of a straight line'. *Can. J. Phys.* 44: 1079-1086.

York, D. 1967. 'The best isochron'. *Earth Planet. Sci. Lett.* 2: 479-482.

York, D. 1969. 'Least squares fitting of a straight line with correlated errors'. *Earth. Planet. Sci. Lett.* 5: 320-324.

York, D., and R.M. Farquhar. 1972. *The Earth's Age and Geochronology*. New York: Pergamon Press.

Zindler, A., and S.R. Hart. 1986. 'Chemical geodynamics'. *Ann. Rev. Earth. Planet. Sci.* 14: 493-571.

Zou, H. 2007. *Quantitative Geochemistry*. London: Imperial College Press.

Index

Abundance sensitivity 83
Acasta gneiss 170
Accelerator mass spectrometer 83, 94
Achondrite 115
Advection 60
Age spectrum 135
Allende meteorite 127
Allende inclusions 133
Alpha (α) decay 171
Anion 143
Antineutrino 13
Ar-Ar dating 137
Archean 169
Atmophile 142, 165
Atom 2
Atomic number 2, 11
Atomic weight 33, 35
Atomic mass unit 3, 34
Auger electron 14
Avogadro number 33

BABI 128
Basalt 154
Basin 184
Basaltic achondrites 132
Batch melting 147
Beryllium 10
Beta decay 13
Big Bang 23, 110
Binomial distribution 19
Binding energy, nuclear 8
Blocking temperature 61
Box model 155
Branched decay 14
Bulk distribution coefficient 145
Bulk silicate earth 76, 142, 171

Ca-Al inclusions 116
Canyon Diablo 128

Carbon burning 27
Carbonaceous chondrite 10, 114
Cation 2
Chain reaction 49
Chalcophile 143
Chart of nuclides 9
Chondrites 115
Chondritic composition 66, 140
Chondritic meteorites 140
Chondrule 115
Chronometer pair 112
CHUR 66, 173
Closed system 67
Closure temperature 58
CNO cycle 26
Compatible element 147
Concordant ages 72
Concordia 71
Condensation 117
Continental crust 156
Convection 152
Cooling history 79
Core 163
Core-mantle boundary 152
Correlation coefficient 104
Correlation diagram 135
Cosmic abundances 9
Cosmic microwave background 23
Cosmic rays 16
Cosmic ray exposure age 138
Collector slit 82
Continental crust 155
Continental growth 169
Cosmochronology 111
Cosmogenic isotopes 57, 138
Covariance 102
Craton 169
Crust 151

Daughter isotope 40, 46
Decay chain 189
Decay constant 40, 128
Decay energy 13
Decay mode 48
Decay series 52
Degassing 142
Delamination 156
Depleted mantle 174
Diamond 154
Differentiation (magma) 79
Diffusion 60
Discordia 71
Distribution coefficient 145
DMM 177
DUPAL anomaly 187
Double focusing 85

Eclogite 155
Electron 2
Electron capture 14
Electron probe micro analyzer 80
Electron volt 3
Electrostatic analyzer 85
EMI 177
EMII 185
Enstatite chondrites 115
e-process 28
Equilibrium partial melting 147
Equiline 190
Errors 97
Error propagation 102
Errorchrons 60
Eucrite 129, 133
Evolution line 176
Exponential law 91
Exposure ages 138
Extended geometry 88
Extinct nuclides 112

Faraday cup 82
Fast neutrons 49
Feldspar 117, 143
Fission products 49
Fission reactor 49
Fission tracks 78
Flood basalt 154
Formation interval 112
FOZO 187

Fractional crystallization 123
Fractionation correction 148
Frequency factor 61
Fundamental forces 3

Ga 31
Gabon 49
Galactic cosmic rays 16, 53, 137
Gamma ray 13, 37
Garnet 154
Geochron 176
Giant impact hypothesis 191
Granite 169
Greenland 170

Hadean 169
Half-life 42
Harzburgite 154
Helium burning 26
Hertsprung-Russell diagram 25
Hf-W systematics 166
Holmes-Houtermans model 71
H-R diagram 25
Hydrogen burning 26
Hydrothermal alteration 183

ICP ion source 88
ICP-MS 90, 182
Igneous rock 143
Incompatible element 147
Induced fission 15
Induced nuclear reaction 15
Initial ratio 52, 173
Internal conversion 13
Inner core 142
Ion 82
Ion detectors 82
Ion microprobe 170
Ionic radius 145
Ionization 77, 86
Ionization potential 86
Iron meteorite 138
Isobars 2
Isochron 58
Isolated system 8, 155
Isotone 2, 7
Isotope 2, 17
Isotope anomalies 188
Isotope dilution 36, 92

Index

Isotope fractionation 34
Isotope homogenization 60
Isotopic equilibrium 63
Isotope evolution diagram 59
Isotopic memory 44
Itsaq Gneiss 170

Ka 43
K-Ar method 51
K-Ca method 51
Kimberlite 154, 177
Kinetic energy 80
Komatiite 169
K-T boundary 184
Krypton 6, 154, 165
Lanthanide 145
Lead paradox 181
Least squares 105
Lherzolite 154
Lithophile 163
Lithosphere 151, 153
Lower crust 179
Lower mantle 187
Lower mantle 152
Lu-Hf system 182
Lunar samples 191

Ma 43
Magma 185
Magma ocean 142, 168
Magnetic sector 94
Main sequence 25
Major element 119
Mantle 151
Mantle array 178
Mantle convection 152
Mantle plumes 152
Mantle wedge 153, 179
Mare 138
Mars 140
Mass defect 8
Mass dispersion 82
Mass discrimination/fractionation 91
Mass number 2, 9
Mass spectrometer 3, 182
Mass-dependent fractionation 162
MC-ICP-MS 90, 182
Mean life 42, 113
Mantle reservoir 185

Meteorite 114
Meteorite falls 116
Mid-ocean ridges 152, 153
Mixing line 76, 179
Mixing time 183
Model age 58
Mole 92
MORB 154
MSWD 60
Multi-stage evolution 61
Mu-meson 17

Nd model age 132, 175
Nebula 11, 112, 193
Negative ions 86
Neutrino 13, 77
Neutron star 29
Noble gases 6, 119
Nuclear binding energy 8
Neutron activation 16
Neutron capture reaction 50, 111
Nuclear reaction 15
Nucleon 2, 9
Nucleocosmochronology 111
Nucleosynthesis 23
Nucleus 1
Nuclide 2, 7, 15

OIB 154, 157, 181
Oceanic lithosphere 153
Oklo nuclear reactor 49
Olivine 116, 154
Open system 71, 155
Ophiolite 154
Orbital 4
Ordinary chondrite 115, 171
Orthopyroxene 154
Osmium 76
Outgassing 166
Oxygen burning 27
Oxygen isotopes 191
Parent isotope 13, 45
Partial melting 123
Partition coefficient 144
Path of stability 7, 12, 31
Pb model ages 132, 133
Per mil 85
Periodic table 5
Peridotite 154, 177

Phosphate 132
Photosphere 9
Plagioclase 116, 150
Planetary accretion 118
Planetesimal 79, 118
Plasma 80, 87
Plate tectonics 152
Plateau age 135
Plutonium 53
Poisson distribution 21
Positron 13
Potential energy 4, 26
Power law 91, 137
p-process 32
Precision 9, 93
PREMA 186
Presolar grain 119
Primordial ratios 128
Probability distribution 18
Proterozoic 169
Proton number 2
Pu-Xe 25
Pyroxene 117

Radioactive decay 17, 58
Radiometric dating 13, 48, 53
Radiation damage 79
Radioactive dating 16
Radioactive disequilibrium 189
Radiocarbon 55
Radiogenic isotopes 44, 162
Radionuclide 8, 41
Rayleigh fractional crystallization 149
Rare Earth Elements (REEs) 6, 145
Rare gases 165
Rayleigh Fractional Melting 148
Rb-Sr system 176
Red giant 27
Red shift 110
Refractory minerals 116
Regression 105
Re-Os system 182
Reproducibility 91
Reservoir age 66
Residence time 139
Rhenium 76
Richardton meteorite 124
Rubidium-strontium (Rb-Sr) method 51, 176

Samarium-neodymium (Sm-Nd) method 51
Scatter diagram 104
Scientific notation 1
Sea floor spreading 152
Seawater 56, 176
Secondary Electron Multiplier (SEM) 83
Secondary ion 87
Secular equilibrium 47, 188
Sedimentary rocks 168, 170
Sedimentation rate 56
SEM 83
Short-lived nuclides 8, 112
SHRIMP 89
Shroud of Turin 55
Siderophile 163
Silicon burning 28
SIMS 89
Single stage evolution 62
Sm-Nd system 173, 182
SMOW 119
SNC achondrites 134
Solar nebula 59, 110, 119
Solar system 9, 110, 141
Solar wind 118, 165
Solidus 144
Spallation 16, 138
Spinel 116
Spontaneous fission 14, 78
s-process 29
Stable isotopes 30, 48
Standard deviation 18, 98
Standard error 99
Steady state 15, 50
Stigmatic focusing 85, 88
Stony iron 115
Stony iron meteorite 115
Strong nuclear force 2
Subducted oceanic crust 179, 186
Subduction zone 153, 179
Sulfur isotopes 34
Supernova 25, 112

T-Tauri phase 165
Tera-Wasserburg diagram 74
Terrestrial residence age 139
Thermal history 76, 136
Thermal ionization 86
Thermal neutrons 15
TIMS 88, 90

Index

Trace elements 143
Trapped electrons 78
Triple alpha process 27
Two-point isochron 66, 170

Ultramafic 116
U-Pb system 71, 128
Upper mantle 76, 150
Uranium isotopes 128
Uranium-Helium (U-He) 134

Vacuum 85
Variance 18
Volcanism 154, 167

White dwarf 25

Xenoliths 154, 176
Xenon 124

Zircon 79, 143